Optimizing and Testing WLANs

Optimizing and Testing WLANs

Proven Techniques for Performance Measurement

By

Tom Alexander

AMSTERDAM • BOSTON • HEIDELBERG • LONDON
NEW YORK • OXFORD • PARIS • SAN DIEGO
SAN FRANCISCO • SINGAPORE • SYDNEY • TOKYO

Newnes is an imprint of Elsevier

Newnes is an imprint of Elsevier
30 Corporate Drive, Suite 400, Burlington, MA 01803, USA
Linacre House, Jordan Hill, Oxford OX2 8DP, UK

∞ Recognizing the importance of preserving what has been written,
Elsevier prints its books on acid-free paper whenever possible.

Library of Congress Cataloging-in-Publication Data
Alexander, Tom.
  Testing 802.11 WLANs : techniques for maximum performance / By Tom Alexander.
      p. cm.
  Includes bibliographical references and index.
  ISBN 978-0-7506-7986-2 (pbk. : alk. paper) 1. Wireless LANs–Security measures.
  2. Local area networks (Computer networks)–Security measures. I. Title.
  TK5105.78.A44 2007
  004.6′8–dc22

                                                                2007017031

British Library Cataloguing-in-Publication Data
A catalogue record for this book is available from the British Library.

ISBN: 978-0-7506-7986-2

For information on all Newnes publications
visit our Web site at www.books.elsevier.com

Typeset by Charon Tec Ltd (A Macmillan Company), Chennai, India
www.charontec.com
Transferred to digital printing in 2009.

Working together to grow
libraries in developing countries
www.elsevier.com | www.bookaid.org | www.sabre.org

ELSEVIER   BOOK AID International   Sabre Foundation

Contents

Preface

My purpose in writing this book is to present a comprehensive review of measurement techniques used in the creation and optimization of IEEE 802.11 wireless LANs. Systematic optimization of a system or process involves extensive measurements, to identify issues and also to know when they have been fixed. A thorough understanding of these measurements and the underlying metrics will aid engineers in improving and extending their wireless LAN equipment and installations.

The extremely rapid development of IEEE 802.11 wireless LANs has resulted in a general lack of usable literature covering their test and measurement. As of this writing, wireless LANs are still in their infancy, and methods of measuring and optimizing their performance are not well understood. In fact, there is much confusion within the industry as to *what* should be measured, let alone *how*. Equipment vendors try to remedy this by publishing articles, whitepapers and application notes, but these are narrowly focused and usually promote the vendor's point of view. It is not unusual to find representatives of leading vendors disagreeing on basic metrics and approaches.

This book tries to present a broad overview of the entire field, to provide the reader with a context and foundation on which more detailed knowledge may be built. My goal is to supply introduction and training material for designers and test engineers. A reader armed with this knowledge should be able to sort out exactly what needs to be measured and how, and what sort of equipment is best suited for the quantity being measured. Such information also allows users, who may not be directly involved in equipment design, to understand the methods that their equipment suppliers *should* have used to measure the numbers claimed on datasheets.

I would like to take this opportunity to thank many colleagues who indirectly contributed to the material covered in this book. In particular, many in-depth discussions of products and test approaches with the employees of VeriWave, Inc. added a great deal to my understanding of the wireless LAN test field. I am especially grateful to Brian Denheyer of VeriWave for a critical review of Chapters 3 and 4, and for making many suggestions for improvement. To my long-review of Chapters 3 and 4, and for making many suggestions for improvement. To my long-suffering editors, Harry Helms and Rachel Roumeliotis, go my heartfelt thanks for their patience and constant encouragement, without which this book might never have been finished. Last but certainly not the least, my gratitude to my wife and family, for unstinting supported throughout.

Tom Alexander

Introduction

The science of metrology is fundamental to all branches of engineering. Before one can engineer a high-performance system, or improve an existing system, one needs to know how to quantitatively measure its performance. After all, if performance cannot be measured in some manner, how will you know if it has improved? In fact, the measurement of physical parameters goes much deeper than performance improvement; in the words of Lord Kelvin, a famous 19th century physicist:

> "In physical science the first essential step in the direction of learning any subject is to find principles of numerical reckoning and practicable methods for measuring some quality connected with it. I often say that when you can measure what you are speaking about, and express it in numbers, you know something about it; but when you cannot measure it, when you cannot express it in numbers, your knowledge is of a meagre and unsatisfactory kind; it may be the beginning of knowledge, but you have scarcely in your thoughts advanced to the state of Science, whatever the matter may be."
>
> – Popular Lectures and Addresses, vol. 1,
> "Electrical Units of Measurement", 1883

The area of test and measurement is therefore a key component of every engineering discipline, and many test instruments provide fascinating examples of engineering ingenuity and precision. Modern microwave test equipment such as spectrum analyzers are often the "hot rods" of the RF world.

This book is devoted to the techniques and equipment used for the test and performance measurement of IEEE 802.11 Wireless LAN (WLAN) devices and systems. It covers test equipment and methods for performance measurements at various network protocol layers: RF (physical), Medium Access Control (MAC), and Transmission Control Protocol/ Internet Protocol (TCP/IP), and application; as well as at various stages: system validation, manufacturing, and installation.

The principal objective of the book is to provide a comprehensive discussion of the performance test problems encountered by wireless engineers, and their solution in the form of measurement systems and procedures. The emphasis is on the underlying engineering principles as well as modern WLAN metrics and methodologies, rather than being a cookbook for technicians. This book is not an encyclopedia of all possible measuring methods; instead, it focuses on specific procedures and setups that are employed in common industry practice. Where viable alternatives exist and are described, their relative merits are also considered. Much of the subject material has been drawn from the author's experience in this field, both as an architect and engineer of WLAN test equipment, as well as a writer of standards for measuring WLAN equipment performance.

Considerable attention has been paid in this book to the difficulties encountered with practical wireless measurement setups, and their solutions. Making useful wireless measurements requires a good understanding of the systematic and equipment errors that can creep into a poorly constructed test setup. Without careful attention paid to such details as signal levels, noise, and isolation, measured results can range from merely irreproducible to completely useless.

This book is therefore aimed at both practicing engineers in many different disciplines, as well as students, engineering managers, equipment reviewers, and even those who are simply curious about how performance figures for WLAN equipment are measured. Engineers dealing with test and measurement functions on a daily basis, of course, form the main audience; the material herein can provide a general background for their work, as well as serving as a reference for specific topics.

As such, engineers specializing in system validation, quality assurance (QA), manufacturing, technical marketing, equipment qualification, WLAN installation, and WLAN maintenance will find useful information presented. For students, managers, and others, it offers an organized introduction to the many different disciplines of WLAN performance measurement, the equipment used, and some understanding of the techniques and complexities of each area. Even design and development engineers, who usually do not run into performance testing on a daily basis, will benefit by knowing how their creations are measured and compared to those from competitors; an in-depth understanding of how a device will be tested is invaluable for understanding how to better design that device.

The material presented in the book is organized as follows:

Chapter 1 provides a brief introduction to IEEE 802.11 WLANs, focusing on the aspects of the various protocol layers that are of interest to people wishing to test them, as well as the architecture and functions of typical WLAN equipment. While readers of this book are expected to be generally familiar with 802.11 technology, it is useful to provide some context and sketch out the general areas of which they are presumed to be aware, in order that they may understand what is to come. However, no attempt is made to provide in-depth coverage of any specific WLAN topic.

Chapter 2 discusses the underlying terminology and concepts of metrology, and covers the different types of test equipment (RF, protocol, installation, etc.) and the various kinds of test processes (design and development, QA, manufacturing, benchmarking, etc.) that are performed by different branches of WLAN engineering. A brief introduction to each area of test and measurement is provided, as well as examples of test setups used in each area; note that these examples should be regarded as merely summarizing the more detailed treatment presented in subsequent chapters. Finally, some common factors affecting the accuracy and validity of WLAN measurements are described.

Chapter 3 treats the different types of environments used to test WLAN equipment (chambers, conducted, over-the-air, etc.), along with their characteristics and limitations. Selection and qualification of a suitable test environment has a significant impact on WLAN test results, and the information presented in this chapter is intended to allow engineers to understand the properties of different types of test environments (e.g., anechoic chambers) as well as to set them up for best results.

Chapter 4 covers physical layer (RF) measurements, focusing principally on the performance characterization required during development and system verification. These tests are usually performed during device-level and board-level verification (i.e., before the complete system is integrated into a final product and manufactured), but may also be carried out as part of system-level performance measurements.

Chapter 5 deals with the diverse measurement methodologies and measuring equipment used to perform WLAN protocol testing. Protocol tests usually cover conformance, performance, and interoperability of complete systems. This area is of most interest to QA and software engineers of WLAN equipment vendors as well as to engineers carrying out qualification and acceptance test procedures on equipment being deployed. Such tests are also used by technical marketing people to compare different brands of equipment, as well as by trade journals to rank vendors' products.

Chapter 6 considers the complicated area of application-level measurements such as voice and video performance, which are of most interest to end-users (and, by extension, the QA and marketing departments of equipment manufacturers). An overview of installed WLAN setups is provided, along with a healthy dose of cautions and caveats, prior to diving into the specifics of measuring the effects of WLANs on voice and video quality.

Chapter 7 covers WLAN manufacturing test, focusing on system-level (rather than chip-level) manufacturing. After a general introduction to WLAN manufacturing processes, some typical manufacturing test setups and equipment are described.

Chapter 8 gives a short introduction to installation (deployment) testing of WLANs in enterprises and hot-spots. The various concerns and issues in WLAN deployment are treated first, as well as the architectures and equipment used in modern WLAN installations. After this, the software and hardware tools and procedures typically encountered while deploying and monitoring WLANs are described. The chapter ends with a discussion of some recent advances in WLAN equipment that can significantly reduce the amount of work and uncertainty involved in WLAN deployment.

Chapter 9 deals with testing IEEE 802.11n systems that employ Multiple Input Multiple Output (MIMO) technology. MIMO is the most recent and exciting development in 802.11 WLANs to date, and both the equipment and the test methods are still under development. The promise of greatly increased bandwidth and resistance to interference of MIMO devices is accompanied by a correspondingly increased measurement complexity. As the field is still in its infancy, the material presented in the chapter goes into rather more depth on the technology and implementation of 802.11n devices, to enable test engineers to understand the new factors that will have to be dealt with when measuring the performance of such systems.

Finally, a pair of appendices are provided, containing references to useful reading material. Appendix A supplies a brief roadmap to the key regulatory and technical standards that govern WLAN engineering; Appendix B contains a bibliography of books and publications that should be consulted for further information.

IEEE 802.11 WLAN Systems

In order to successfully test something, it is essential to have a good understanding of how it works and what it does. We will therefore begin with an introduction to the important technical factors behind IEEE 802.11 wireless LANs (WLANs), as well as the standards and regulatory documents that govern how WLANs are developed and operated. By necessity, only brief explanations can be provided here; the reader is encouraged to consult the actual standards documents and other references for more information.

1.1 IEEE 802.11 Wireless Local Area Networks

Contrary to popular misconception, 802.11 is not merely "wireless Ethernet." Instead, 802.11 WLANs use an entirely different network protocol and are deployed in different topologies. The purpose of a WLAN is primarily to provide LAN connectivity to portable and mobile stations (laptop computers, voice handsets, bar-code readers, etc.), though fixed-station use is becoming more popular as the technology becomes widely adopted.

Essentially, WLANs provide data communications over radio links, and are subject to all the vagaries of RF propagation and interference that any radio communications system suffers. Wired (optical or copper) LAN links are nearly error-free (normal bit error rates are on the order of 1×10^{-9}), physically secure, independent of environmental influences or mutual interference, and provide extremely high bandwidth. A single optical fiber, for instance, is capable of supporting hundreds of gigabits/second of bandwidth. By contrast, radio links are subject to error rates as high as 10%, subject to both eavesdropping and denial of service, highly affected by propagation characteristics and nearby equipment, and support only 10–500 Mb/s of bandwidth that must be shared between all users of the RF channel. As radio signals propagate well outside the area covered by the WLAN and could interfere with other radio services, the operation of WLANs is governed by national and international regulations rather than being exclusively limited by technical or market considerations. The following table summarizes the key differences between wired (optical or copper) and wireless LANs.

Attribute	Wired LANs	Wireless LANs
Data rates (2006)	10 Mb/s–10 Gb/s	1–54 Mb/s
MAC protocol	CSMA/CD(Carrier Sense Multiple Access/Collision Detection)	CSMA/CA (Carrier Sense Multiple Access/ Collision Avoidance)
Range	500 m or more	50 m or less
Error rates	1×10^{-9} to 1×10^{-12}	1×10^{-5}
Usage	Throughout the enterprise	Access links to wired infrastructure
Mobility	None	Mobile
Medium access	Typically switched (each user has a separate channel)	Typically shared (many users share a common channel)
Operating mode	Connectionless	Connection oriented
Interference	Nearly non-existent	Highly susceptible
Affected by environment	Almost completely independent of surrounding environment	Highly affected by RF propagation characteristics of environment
Physical security	Easy to provide	Requires advanced encryption
Implementation complexity	Relatively low	Highly complex
Devices connected	Computers, switches, routers	Computers, switches, laptops, personal digital assistants (PDAs), phones, bar-code scanners, RFID tags, etc.

While the IEEE 802.11 protocol allows for different types of WLAN topologies to be set up, nearly all deployed WLANs comprise two types of stations: clients and access points (APs). Clients such as laptops are the endpoints in the WLAN, and run the applications that source and sink data traffic. APs, on the other hand, provide portals into the remainder of the wired LAN; it is rare to find a LAN that is exclusively comprised of wireless devices. They support wireless interfaces on the "front" and wired interfaces such as Ethernet, DSL, or DOCSIS cable at the "back", and act as bridges between the wired and wireless infrastructure. Clients associate (connect) with APs to exchange data traffic with each other or the remainder of the LAN or WAN.

A group of clients and APs is collectively referred to as a service set. The 802.11 standard defines two kinds of service sets: a basic service set (BSS), which comprises a single AP and some number of clients; and an extended service set (ESS), which joins together several APs into a common network by means of a wired infrastructure. We will be concerned principally with ESS network operations in this book.

The following figure depicts the reference model under which 802.11 WLANs operate.

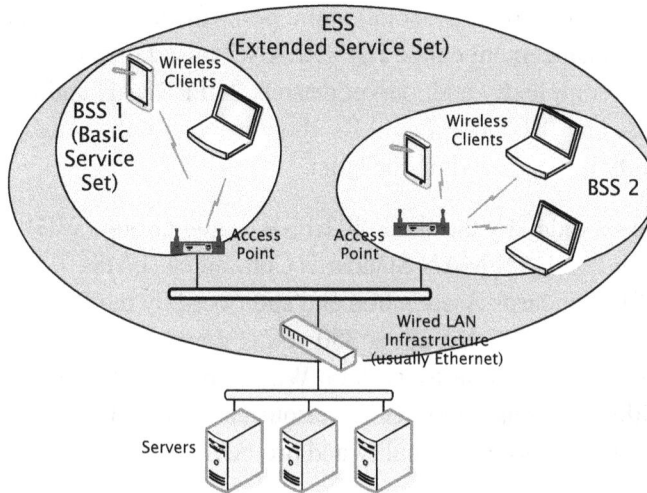

Figure 1.1: The 802.11 Reference Model

It is plain from the above figure that the wireless data links of WLANs coexist with wired Ethernet links. WLANs normally replace the "last 30 feet" of a data communications network to provide mobility, but are not used in the remainder of the network, where the emphasis is on bandwidth (large servers and routers, after all, do not move about). Data traffic carried over WLAN links uses the Transmission Control Protocol (TCP)/Internet Protocol (IP).

1.2 WLAN Standards Today

In 1985, the Federal Communications Commission (FCC) decided to open up the so-called ISM (Industrial, Scientific, and Medical) bands for use by unlicensed low-power communication devices using spread-spectrum modulation methods. This spurred significant interest in the US in developing wireless networking equipment utilizing these bands for computer communications (i.e., radio LANs) to serve as a radio version of the popular Ethernet LAN technology. As a result, in 1990 the IEEE standards development organization set up a group, referred to as the IEEE 802.11 committee, to standardize WLANs in the ISM bands. However, it took 7 years (until 1997) before the first 802.11 standard was ratified and published. That first standard defined a relatively low-speed digital WLAN technology, with data rate options of 1 and 2 Mb/s, and using a new Carrier Sense Multiple Access/Collision Avoidance (CSMA/CA) medium access protocol, which was roughly modeled after the Carrier Sense Multiple Access/Collision Detection (CSMA/CD) protocol used by half-duplex IEEE 802.3 (Ethernet) LANs.

In parallel with the work of the IEEE committee, the European Telecommunications Standards Institute (ETSI) started work in 1991 on a radio LAN technology called HIPERLAN (High Performance European Radio LAN). HIPERLAN was standardized somewhat earlier than

IEEE 802.11 (1996) and offered considerably more performance: 10 Mb/s, as compared to 2 Mb/s. A subsequent enhancement called HIPERLAN/2 raised this to 54 Mb/s in the year 2000. However, due to complexity and market reasons, HIPERLAN and HIPERLAN/2 have been largely superseded by IEEE 802.11 LANs, though some of the principles of the former have been subsequently incorporated by the latter.

WLAN standards are set today by the IEEE 802.11 Working Group (WG), which is a subsection of the IEEE 802 LAN/MAN Standards Committee (LMSC), which in turn is a subsection of the IEEE Standards Association and sponsored by the IEEE Computer Society. As of this writing, the 802.11 WG has about 350 voting members and several hundred observers, and meets six times a year to work on WLAN-related standards. The 802.11 committee works within the constraints set by various national and international regulatory bodies to define the actual radio functionality and protocol.

The IEEE 802.11 standard does not try to specify how a WLAN device should be constructed – it leaves the design and operation of the actual clients and APs up to the implementer. Instead, it specifies the interactions between WLAN devices, collectively referred to as the WLAN protocol. The purpose of the standard is to ensure interoperability between devices without unduly constraining the device designer or vendor.

The WLAN protocol is partitioned into a number of pieces or layers:

1. The physical or PHY layer, which deals with the transmission and reception of radio signals, and is further divided into the physical media-dependent (PMD) portion and the PHY-layer convergence protocol (PLCP).

2. The Medium Access Control or MAC layer, which deals with the exchange of suitably formatted packets.

3. The PHY management layer, which handles the interactions required to control the PHY layer.

4. The MAC management layer, which likewise deals with the interactions needed to control the MAC layer.

The 802.11 WLAN standard is thus actually a collection of related standards, specifying all of the pieces described above. To date, there are over 25 different protocols and subprotocols comprising the 802.11 protocol stack, each being created (or having been created) by a separate subgroup within IEEE 802.11. The following figure shows a rough map of this plethora of protocol elements. The reader should observe the caveat that, as with any dynamic standards body, the number of protocols grows by leaps and bounds every year.

IEEE 802.11 subgroups are known as Task Groups (TGs), and are assigned letter suffixes to distinguish one from the other. The standards documents that they create are also assigned

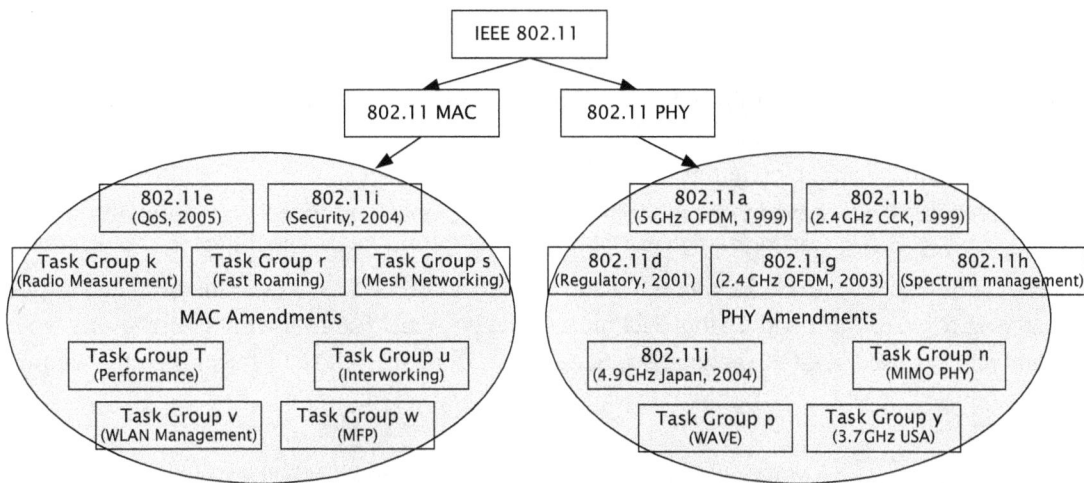

Figure 1.2: A Zoo of Protocols

these same letter suffixes. For example, TGg created a PHY layer standard for Orthogonal Frequency Division Multiplexing (OFDM) transmission in the 2.4 GHz band, which promptly became known as 802.11 g. Similarly, TGi introduced a much enhanced security system, which was enshrined in the 802.11i standards document (more commonly known as WPA2, after the Wi-Fi® Alliance nomenclature). A curious convention is used when assigning letter suffixes: lowercase letters denote standards documents that will eventually be folded into the main 802.11 standard, while uppercase letters indicate that the document will remain permanently stand-alone. Thus the output of the 802.11b group was folded into the main 802.11 document in 2003 (forming Clause 18), but the 802.11T group is creating the 802.11.2 document, which will remain as a stand-alone performance test specification.[1]

1.2.1 PHY Standards

In the US, the PHY layer of 802.11 occupies two principal microwave frequency bands: the ISM band at 2.400–2.483 GHz, and the Unlicensed National Information Infrastructure (U-NII) band at 5.150–5.825 GHz. (There is a further allocation in the 4.900 GHz public service band, but this is a relatively recent development.) All 802.11 WLANs share these frequency ranges with other users, most notably microwave ovens in the 2.4 GHz band. In theory, as 802.11 WLANs only have a secondary allocation in these bands, a WLAN must cease operation if it causes interference to the primary users; in practice, however, this almost never happens, due to the low power used by 802.11 radios.

---

[1] 802 standards are copyrighted by the IEEE. All 802.11 standards are available for on-line download at www. getieee802.org, or may be ordered in electronic or paper form directly from the IEEE.

The original 802.11 standard called for a 2.4 GHz time-division-duplex (TDD) radio link with data rates of 1 and 2 Mb/s, using DBPSK and DQPSK modulation, respectively. Both direct-sequence spread-spectrum (DSSS) and frequency-hopping spread-spectrum (FHSS) methods were specified and deployed; TDD was used to allow the uplink and downlink signals to share the same channel, taking turns to transmit. While FHSS was generally more robust to interference, DSSS proved to be more efficient and flexible, and FHSS was gradually abandoned; no vendor sells 802.11 FHSS radios today. Subsequently, the 802.11b standard added Complementary Code Keying (CCK) at 5.5 and 11 Mb/s data rates to the mix, in addition to carrying forward the 1 and 2 Mb/s data rates of the original. The following figure shows the general process used in CCK modulation. See Clause 18 of IEEE 802.11 for more information.

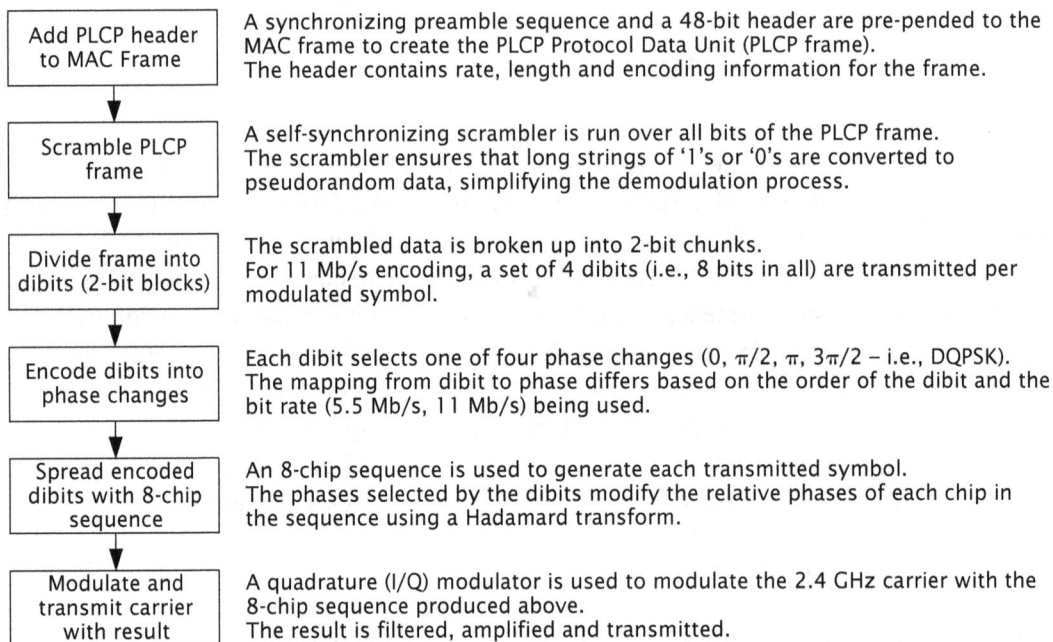

Add PLCP header to MAC Frame	A synchronizing preamble sequence and a 48-bit header are pre-pended to the MAC frame to create the PLCP Protocol Data Unit (PLCP frame). The header contains rate, length and encoding information for the frame.
Scramble PLCP frame	A self-synchronizing scrambler is run over all bits of the PLCP frame. The scrambler ensures that long strings of '1's or '0's are converted to pseudorandom data, simplifying the demodulation process.
Divide frame into dibits (2-bit blocks)	The scrambled data is broken up into 2-bit chunks. For 11 Mb/s encoding, a set of 4 dibits (i.e., 8 bits in all) are transmitted per modulated symbol.
Encode dibits into phase changes	Each dibit selects one of four phase changes ($0, \pi/2, \pi, 3\pi/2$ – i.e., DQPSK). The mapping from dibit to phase differs based on the order of the dibit and the bit rate (5.5 Mb/s, 11 Mb/s) being used.
Spread encoded dibits with 8-chip sequence	An 8-chip sequence is used to generate each transmitted symbol. The phases selected by the dibits modify the relative phases of each chip in the sequence using a Hadamard transform.
Modulate and transmit carrier with result	A quadrature (I/Q) modulator is used to modulate the 2.4 GHz carrier with the 8-chip sequence produced above. The result is filtered, amplified and transmitted.

Figure 1.3: CCK Modulation Process

The data exchanged between 802.11 stations, at the PHY layer, is encapsulated within a frame format known as the PLCP frame. PLCP frames are different for the various modulation schemes, but generally contain a short header that indicates the coding and length of the encapsulated MAC frame; the receiver then uses this to properly decode the frame. The PLCP frame transmitted by an 802.11b radio is shown in the figure below.

The 802.11a standard was approved after the adoption of the 802.11b standard. (Actually, work on the 802.11a standard was started prior to 802.11b, but as it used a much more

Sync (Scrambled Ones) (128 bits)	SFD (16 bits)	Signal (8 bits)	Service (8 bits)	Length (16 bits)	CRC (16 bits)	MAC	Frame

Long PLCP Preamble (144 bits at 1 Mb/s) Long PLCP Header (48 bits at 1 Mb/s)

Sync (Scrambled Zeros) (128 bits)	SFD (16 bits)	Signal (8 bits)	Service (8 bits)	Length (16 bits)	CRC (16 bits)	MAC	Frame

Short PLCP Preamble (72 bits at 1 Mb/s) Short PLCP Header (48 bits at 2 Mb/s)

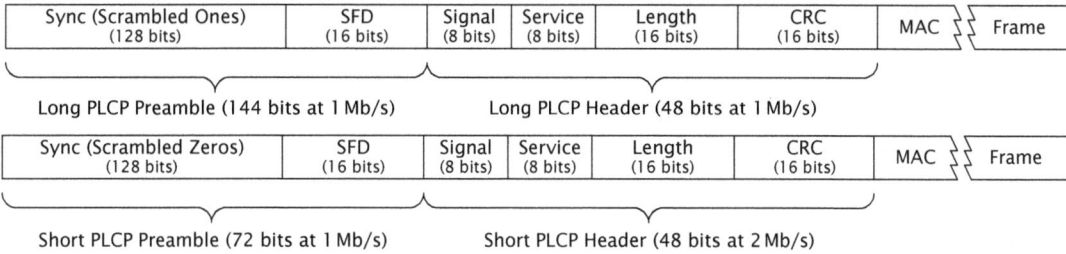

Figure 1.4: 802.11b PLCP Frame

complex modulation scheme – OFDM – it took longer to develop than 802.11b. Hence the puzzling inversion in the nomenclature.) The 802.11a standard operates in the 5.8 GHz band, and calls for several different modulation types to achieve a large range of PHY bit rates. The modulation types are not only the BPSK and QPSK used in the 1 Mb/s PHY, but also include 16-QAM (quadrature amplitude modulation) and 64-QAM, leading to much higher data rates: 6, 9, 12, 18, 24, 36, 48, and 54 Mb/s. These modulation types are imposed on a set of 52 subcarriers spread over a 16.6 MHz channel bandwidth. A block diagram of the OFDM modulation and transmission process is shown below; Clause 17 of IEEE 802.11 provides details.

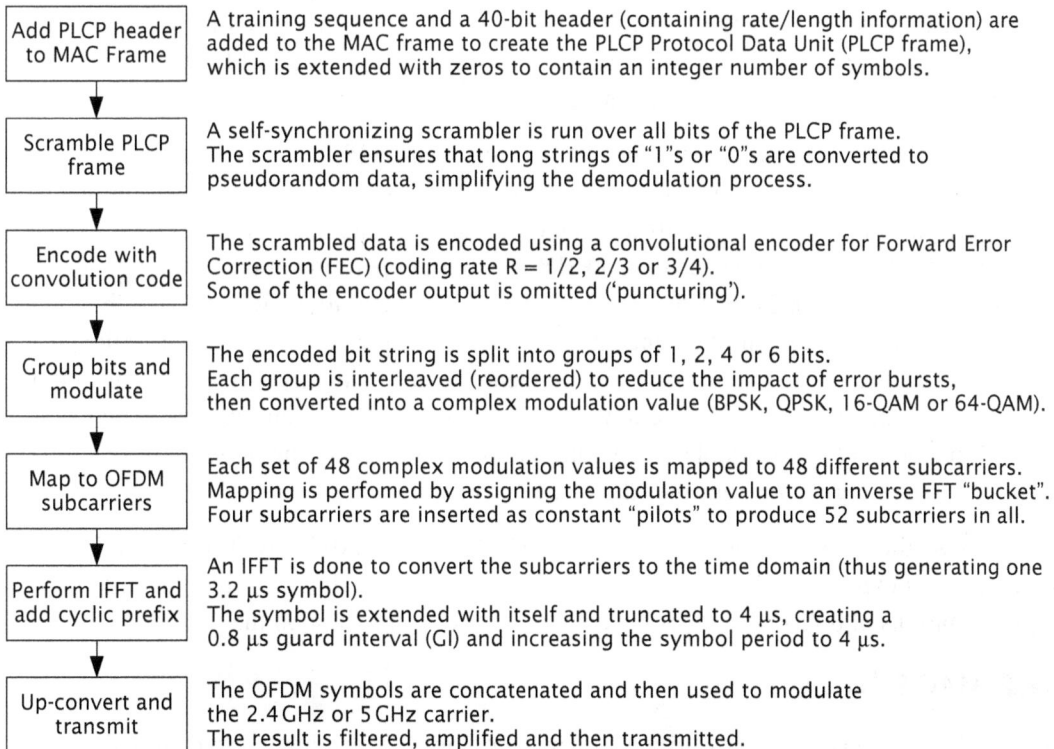

Add PLCP header to MAC Frame	A training sequence and a 40-bit header (containing rate/length information) are added to the MAC frame to create the PLCP Protocol Data Unit (PLCP frame), which is extended with zeros to contain an integer number of symbols.
Scramble PLCP frame	A self-synchronizing scrambler is run over all bits of the PLCP frame. The scrambler ensures that long strings of "1"s or "0"s are converted to pseudorandom data, simplifying the demodulation process.
Encode with convolution code	The scrambled data is encoded using a convolutional encoder for Forward Error Correction (FEC) (coding rate R = 1/2, 2/3 or 3/4). Some of the encoder output is omitted ('puncturing').
Group bits and modulate	The encoded bit string is split into groups of 1, 2, 4 or 6 bits. Each group is interleaved (reordered) to reduce the impact of error bursts, then converted into a complex modulation value (BPSK, QPSK, 16-QAM or 64-QAM).
Map to OFDM subcarriers	Each set of 48 complex modulation values is mapped to 48 different subcarriers. Mapping is performed by assigning the modulation value to an inverse FFT "bucket". Four subcarriers are inserted as constant "pilots" to produce 52 subcarriers in all.
Perform IFFT and add cyclic prefix	An IFFT is done to convert the subcarriers to the time domain (thus generating one 3.2 μs symbol). The symbol is extended with itself and truncated to 4 μs, creating a 0.8 μs guard interval (GI) and increasing the symbol period to 4 μs.
Up-convert and transmit	The OFDM symbols are concatenated and then used to modulate the 2.4 GHz or 5 GHz carrier. The result is filtered, amplified and then transmitted.

Figure 1.5: OFDM Modulation Process

The 802.11a PLCP frame is different from the 802.11b frame, and is shown below.

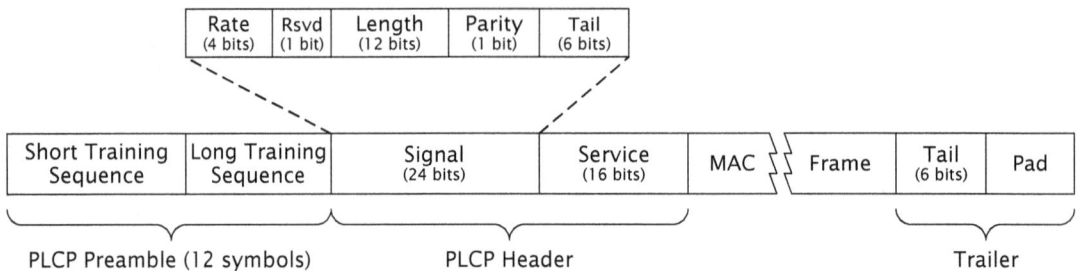

Rate (4 bits)	Rsvd (1 bit)	Length (12 bits)	Parity (1 bit)	Tail (6 bits)

Short Training Sequence	Long Training Sequence	Signal (24 bits)	Service (16 bits)	MAC	Frame	Tail (6 bits)	Pad

PLCP Preamble (12 symbols)   PLCP Header   Trailer

Figure 1.6: 802.11a PLCP Frame

The 802.11a PHY operates in the 5.15–5.825 GHz band, which suffers from indoor propagation limitations. Due to market demand, therefore, the 802.11 WG began work on extending these same data rates to the 2.4 GHz band shortly after 802.11a was published. The result was the 802.11g standard, which incorporated all of 802.11b for backwards compatibility, and added the OFDM modulation types from 802.11a as well, producing a plethora of data rates: 1, 2, 5.5, 6, 9, 11, 12, 18, 24, 36, 48, and 54 Mb/s. (The specific data rate to be used is selected by the transmitter according to the channel conditions, to assure the best chance of getting the data across in the shortest time.) The 802.11g standard remains today the most widely used WLAN physical layer.

In 2004, work was started within the 802.11 WG to specify a PHY that utilized the substantial bandwidth gains available when using multiple antennas, a technique known as Multiple Input Multiple Output (MIMO). This led to the formation of the 802.11n task group, which is currently in the process of specifying a PHY capable of operating at data rates between 6.5 and 600 Mb/s in both 2.4 and 5 GHz bands. The MIMO technique will be described in some more detail later, but in essence it uses several independently driven transmit and receive antennas to create two or more independent "virtual" streams between a transmitter and a receiver, and then sends different blocks of data down the various streams. The result is a multiplication of the available bandwidth without a corresponding increase in spectrum occupancy. The figure below outlines the MIMO concept.

As of this writing, the work on standardizing 802.11n is still under way. The final 802.11n standard is not expected to be ratified until 2008 at the earliest, though "pre-standard" implementations of 802.11n devices have already begun appearing on the market.

1.2.2 MAC Sublayers

The 802.11 MAC layer is necessarily a somewhat complex beast, having to deal with the vagaries of TDD radio links and mobile users. (To illustrate this: while the formal description

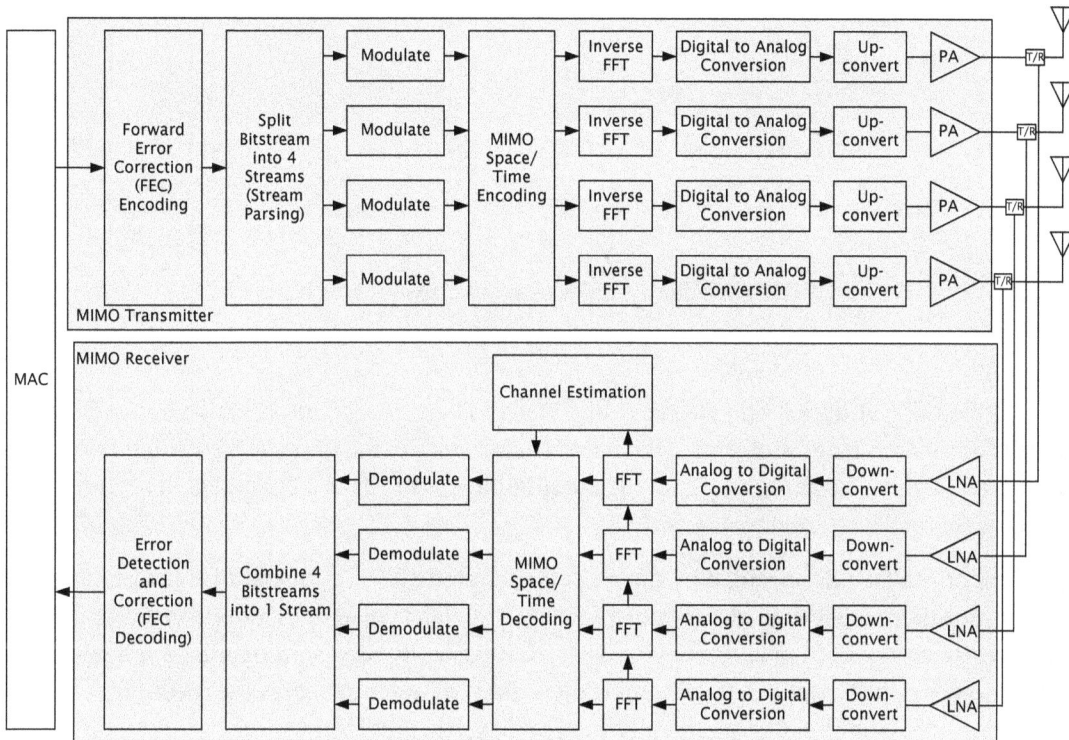

Figure 1.7: MIMO PHY

of the entire 802.3 Ethernet MAC layer requires barely 15 pages, in comparison, the formal description of the 802.11 MAC extends to over 200!) It is also blessed with no less than four different operating modes, of which two are closely related and actually used in common practice.

The most common 802.11 MAC operating mode is referred to rather obscurely as the Distributed Coordination Function (DCF), and is specified in subclause 9.2 of IEEE 802.11. The DCF is a variant on the CSMA/CD half-duplex access method employed in Ethernet; stations always listen before transmitting, and hold off (defer) to transmissions that have started earlier. If two stations happen to transmit simultaneously, the result is a collision, and neither station will be successful. In Time Division Duplex (TDD) radio links, however, it is not possible to directly detect a collision, as the receiver is usually shut off (muted) during transmit to avoid being overloaded. Instead, an indirect collision sensing scheme is used: every transmitted packet is acknowledged, and the lack of an acknowledge indicates that the packet was not successfully received, and should be retransmitted. This has the additional benefit of automatically handling the high frame error ratio of radio links – errored frames are simply retransmitted.

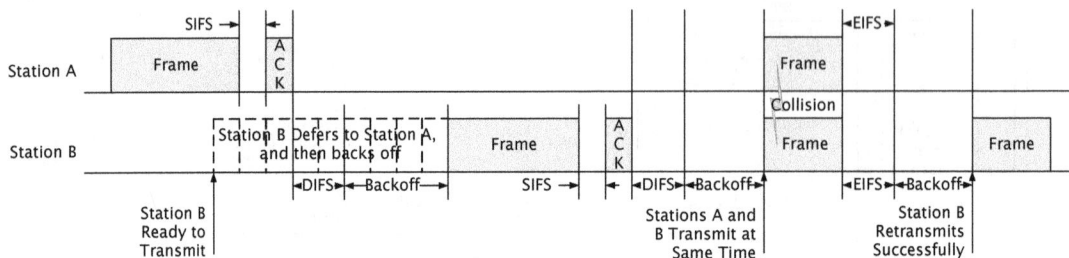

Figure 1.8: DCF Medium Access (see subclause 9.2.5, IEEE 802.11)

Further, the DCF utilizes a scheme for collision avoidance, forcing prospective transmitters to wait for random lengths of time – the backoff interval – in the hope of preventing two transmitters from attempting to get on the air simultaneously. The access method used by 802.11 is therefore referred to as CSMA/CA.

A variant of the DCF is specified by the recently adopted 802.11e standard for prioritizing medium access for real-time, delay-sensitive traffic such as voice or video. Referred to as Enhanced Distributed Channel Access or EDCA, it basically uses a probabilistic scheme, forcing lower priority stations to wait for longer times in order to access the medium, while higher priority stations suffer a generally lower delay. This results in voice or video traffic obtaining preferential access to the wireless medium, while data traffic takes what bandwidth is left.

The two other operating modes are referred to as the Point Coordination Function (PCF) and Hybrid Coordination Function (HCF) Controlled Channel Access (HCCA). The PCF is a centralized, polling-based access method, where the AP is responsible for controlling which stations are permitted to transmit, and polling all stations using special control packets to determine if they need to send data. HCCA is the QoS variant of PCF, and defined in 802.11e. Neither are commonly used in operating WLANs today – in fact, the author is not aware of any equipment that even implements PCF – and so will not be described further.

In addition to the basic channel access functions, the 802.11 standard encompasses a number of extensions and additional protocols for security, QoS support, radio channel and neighbor station assessment, roaming, etc. The original security method provided for by 802.11 was the infamous WEP (Wired Equivalent Privacy) protocol, which relied on fixed, manually configured encryption keys for the RC4 encryption protocol. The 802.11i standard rectified three of the biggest flaws of WEP – weak encryption keys, manual configuration, and lack of protection against replay attacks – with a much more comprehensive scheme utilizing the IEEE 802.1X protocol for dynamic generation and distribution of encryption keys. Similarly, the 802.11e standard added QoS functions to 802.11 networks. In addition to defining the EDCA and HCCA prioritized medium access methods, the 802.11e standard provided mechanisms to perform admission control (i.e., preventing the network from being overloaded) and traffic management. Other 802.11 letter suffixes (802.11k, 802.11r, etc.) add even more capabilities to the base MAC standard.

1.2.3 Other Related Standards

People involved with the technical aspects of 802.11 devices and systems usually have to familiarize themselves with a small collection of related standards documents as well. The most obvious one, of course, is the IEEE 802.3 (Ethernet) standard; virtually every AP or wireless gateway has at least one Ethernet port, sometimes more, incorporated. In fact, before the advent of residential wireless gateways that integrated DSL or cable modems, the sole function of a wireless AP was to bridge WLAN traffic to an Ethernet LAN.

The location of Ethernet devices in a WLAN topology is the same regardless of whether the WLAN is being used in a residence or a corporate environment: the Ethernet LAN sits between the WLAN and either the Internet or the corporate WAN connection. The Ethernet LAN serves to link together some number of APs, the servers or routers that supply data services required by the wireless clients, and the WAN interface. In some cases the Ethernet LAN even facilitates WLAN-specific functions; for example, the pre-authentication protocol specified by 802.11i for fast wireless roaming applications is actually performed over the Ethernet network.

As the Ethernet frame format is quite different from the 802.11 frame format, the AP performs a frame translation process during the bridging of data between Ethernet and 802.11. (The frame translation causes the frame to grow or shrink in size, causing quite a bit of confusion when interpreting the results of traffic throughput tests – but more about that later.) The 802.11 frame contains extra address fields to enable the AP to construct a valid Ethernet frame and direct it to the appropriate destination. Most of the 802.11-specific information, however, does not make it across the AP's interface. Thus a packet "sniffer" sited on an Ethernet LAN will not be able to see any of the 802.11 control or management frames, or 802.11-related information in data frames.

The other standard that is intimately tied up with 802.11 WLANs is TCP/IP, which is the only higher-layer protocol that 802.11 is currently defined to support. Both TCP and IP are standardized by the Internet Engineering Task Force (IETF); their formal definitions may be found in Request For Comment (RFC) 793 and RFC 791, respectively.

In enterprise WLANs employing centralized (server-based) security, another protocol is often used: IEEE 802.1X, also known as EAPOL. The 802.1X specifies a transport mechanism for passing various kinds of authentication packets between 802.11 clients and a security server, typically one that runs the RADIUS (Remote Authentication Dial-In User Service) protocol, which allows the clients to establish authentication credentials (usernames, passwords, certificates, etc.) in a secure manner. The actual authentication exchanges between the client and the server commonly follow the Extensible Authentication Protocol (EAP) defined in RFC 3748 – hence the acronym EAPOL for 802.1X, which stands for EAP Over LANs. The whole area of security is crucial to the setup and operation of modern enterprise WLAN devices; the reader is referred to the book *Real 802.11 Security* by Edney and Arbaugh for a good introductory explanation of the subject.

Finally, the centralization of AP management and configuration is becoming quite a significant trend in enterprise WLANs. Enterprise WLAN vendors have been adopting a model where most or all of the configuration functions are automatically performed on all the APs by a central box referred to as a WLAN switch. (Basically the network administrator configures the WLAN switch, and it in turn configures all the APs over the wired LAN.) The protocol between the WLAN switch and the APs has usually been proprietary and closed, but a new IETF WG – CAPWAP, standing for Control and Provisioning of Wireless Access Points – has been working on a standardized protocol for this purpose.

1.2.4 Regulatory Bodies and Standards

Wired or optical networking technologies usually exist purely under the control of vendors and users, unfettered by governmental rules and regulations. As previously mentioned, WLANs are different: they use radio spectrum which is managed by international treaty at the World Administrative Radio Conference (WARC) and the International Telecommunications Union (ITU), and are therefore subject to regulations set up by independent government-appointed regulatory bodies in various countries. Each country (or administrative region, such as the European Economic Community (EEC)) promulgates its own set of regulations that WLANs need to follow in order to be allowed to operate.

In the US, WLAN regulation is performed by the FCC, under Part 15 of Title 47 of the Code of Federal Regulations. The FCC sets the radio channels that can be used, the maximum power output of the transmitters, and the basic modulation characteristics. Other countries, of course, have their own rules and regulations. The following table summarizes the principal regulatory bodies and rules.

Country	Authority	Rules Documents
United States	FCC	CFR Title 47 Part 15
Canada	Industry Canada	GL-36
Europe	ETSI National PTTs	ETS 300
Japan	Ministry of Telecommunications (MKK)	RCR STD-33 published by Association of Radio Industries & Businesses (ARIB)

1.3 Inside WLAN Devices

This section briefly describes the "guts" of various WLAN devices. In order to test a device, it is necessary to have at least some basic understanding of how the device works and what is inside it. The description is necessarily fairly superficial; the reader is referred to datasheets and product descriptions for more information. (In some cases, even product literature will not help; there is no substitute for taking a device apart to see what makes it tick.)

1.3.1 Clients

Clients are at the base of the WLAN pyramid, and are the only elements that are actually in the hands of users. WLAN clients comprise basically any device that has a wireless interface and actually terminates (i.e., sources or sinks) data traffic. Examples of devices that can act as WLAN clients are: laptops (virtually every laptop shipped today contains a WLAN interface), PDAs, VoIP telephone handsets, game consoles, bar-code readers, medical monitoring instruments, point-of-sale (POS) terminals, audiovisual entertainment devices, etc. The number of applications into which WLANs are penetrating grows on a monthly basis; the WLAN toaster is probably not too far in the future!

The WLAN portion of a client is required to perform the following functions:

1. Association (connection) with a counterpart device, such as an AP. (Prior to association, the client is not permitted to transfer any data.)

2. Security and authentication functions to assure the counterpart device that the client is in fact who it says it is, and is authorized to connect.

3. Protocol stack support, principally of the TCP/IP protocol, so that applications can transfer data once the connection process is completed and everything is authorized.

4. Mobility functions, such as scanning for higher-power APs and "roaming" from AP to AP when the client is in motion.

The counterpart device to which a WLAN client connects is almost always an AP. The 802.11 protocol standard does allow a client to connect directly to another client (this is referred to as "ad hoc" mode), but this mode is almost never used; in fact, ad hoc mode represents a management and security headache for most IT staff.

A "typical" client (insofar as there can be a typical client) comprises two elements: a hardware network interface card or module, and a large assemblage of firmware and software. The following figure depicts the general architecture of a client.

The network interface card is typically a PCMCIA (PC-Card) or mini-PCI card for a laptop or PDA, or may be built into an integrated module in the case of phones or bar-code readers. The level of silicon integration for WLAN NICs is extremely high. In the most highly

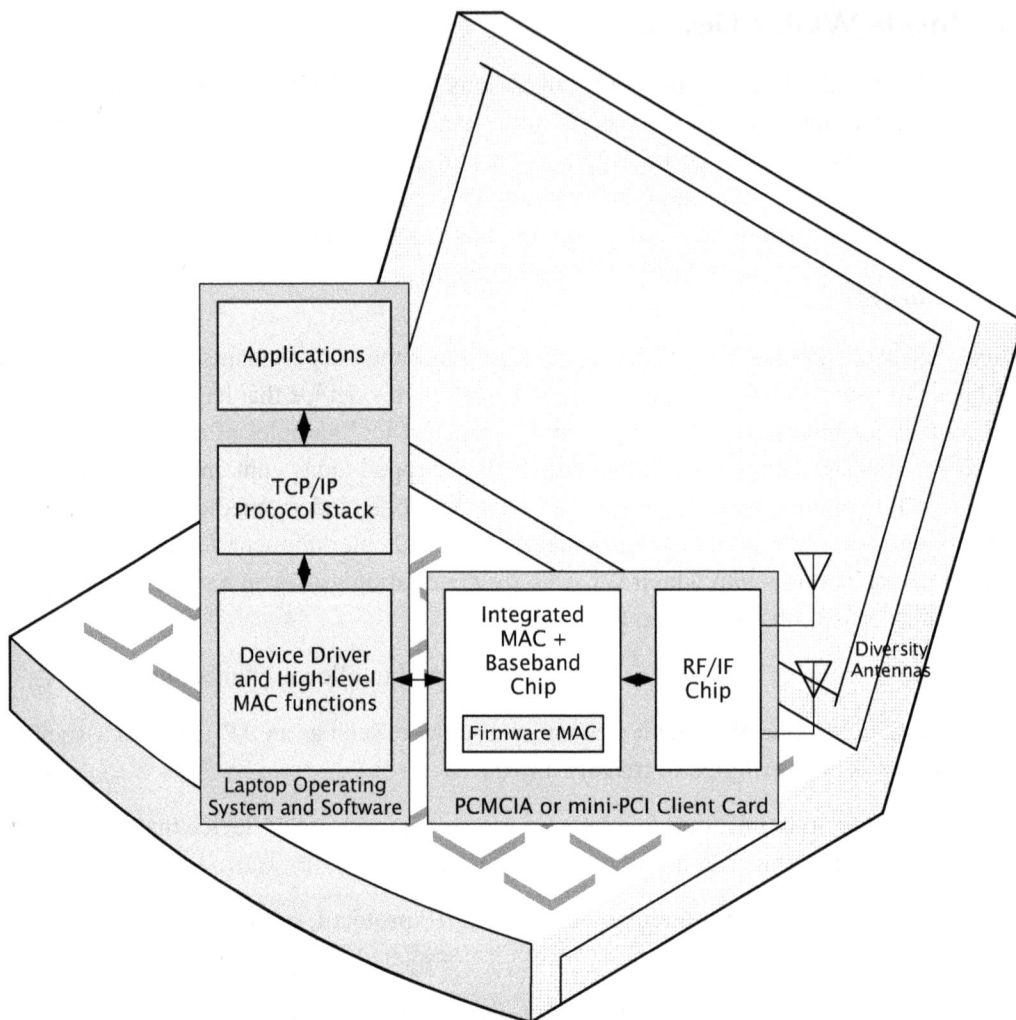

Figure 1.9: A Typical Client

integrated form, a NIC may consist simply of a single CMOS chip supporting the RF and IF functions (up/down conversion, amplification, frequency synthesis and automatic gain control or AGC), the baseband functions (modulation and demodulation, and digitization), and the lower layers of the MAC functions (packet formatting, acknowledgements, etc.). In this case, external passive and small active parts are all that is necessary to create a complete NIC. More commonly, an NIC can comprise two devices: a fully digital MAC and baseband chip, usually fabricated in CMOS, and a separate RF/IF device that may be fabricated using silicon–germanium (SiGe) or other high-speed technology. Note that most NICs today support operation in both 2.4 GHz and 5.8 GHz frequency bands (not at the same time) and contain two separate RF/IF chains, one for each frequency band. The chains are frequently integrated into a single SiGe device, though.

The silicon portion of a client normally only performs the lowest layer of the MAC functions: packet formatting, checking, encryption/decryption, acknowledgements, retransmissions, and protocol timing. The remainder of the MAC functions – typically referred to as the upper MAC – comprise authentication/association, channel scanning, power management, PHY rate adaptation, security, and roaming. These are almost always implemented using a combination of firmware, device drivers, and operating system software. (Many MAC chips integrate a small ARM or MIPS RISC processor to support some of the firmware functions.) In the case of laptops or Windows CE PDAs, the Windows OS performs a good portion of the upper-layer 802.11 functions. In general, the partitioning of functions is done as follows: low-level, real-time tasks are done by the hardware, mid-level protocol functions by the firmware or device driver, and higher-level, user-visible tasks (such as selecting a specific network to associate with) are carried out by the operating system and the WLAN card management processes running under it.

1.3.2 Access Points

APs form the essential counterpart to clients in almost every modern WLAN. APs comprise exactly what their name implies: they provide points at which clients can gain access to the wired infrastructure, bridging between the wireless (RF) world and the Ethernet domain. While in a home environment the number of APs may almost equal the number of clients (it is not unusual to find home WLANs consisting of exactly one client and one AP), in typical enterprise installations the clients outnumber the APs by a factor of 5 or more. Enterprise equipment vendors usually recommend that no more than 6 to 10 clients be supported by each AP.

The functions of an AP are in many cases a mirror image of those performed by a client:

1. Broadcasting "beacons" to indicate their presence and abilities, so that clients can scan for and find them.

2. Supporting association by clients, as well as the security handshakes required by whatever security scheme is being used. Note that APs do not actually process any of the security handshakes apart from the ones defined by the 802.11 and 802.11i standards; instead, they establish a secure connection to a RADIUS server and pass these packets on.

3. Bridging and packet translation of data packets sent to or received from connected clients.

4. Buffering of packets, especially in the case of "sleeping" clients that are using the 802.11 power management protocol to conserve battery life.

In many cases, APs also participate in "RF layer management", especially in large enterprise deployments. In this case, they monitor for adjacent APs, detect "rogue" APs and clients, adjust their signal strength to limit interference, and pass information up and down the protocol stack to enable clients to roam quickly.

The following figure shows the typical internal architecture of an enterprise-class AP.

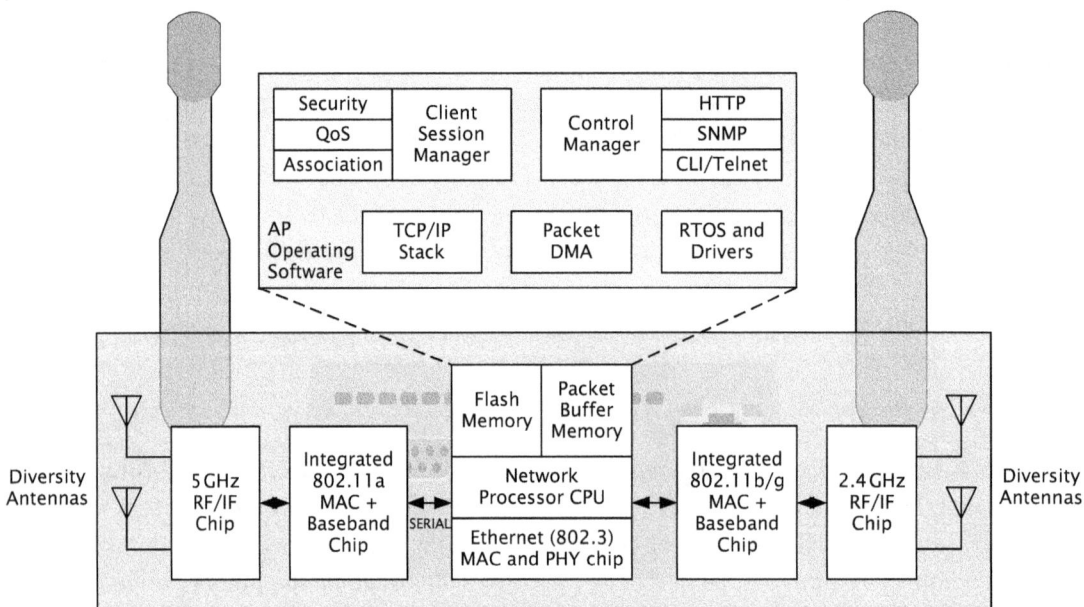

Figure 1.10: A Typical Access Point

The hardware portion of the AP is not unlike that of a laptop client, comprising a device to perform RF/IF functions and another, more integrated device that contains the MAC and baseband functions. However, there are two key differences:

1. Many APs (enterprise APs in particular) support simultaneous operation in both the 2.4 and 5.8 GHz frequency bands. Thus they contain two completely independent RF/IF chains, basebands, and MAC processing elements.

2. Client NICs can rely on the presence of a host CPU and OS, but APs cannot. Thus APs typically integrate some kind of control CPU running an embedded OS (frequently some version of Linux) for these functions.

The firmware functions in an AP are, however, entirely different. The need to support the 802.11 protocol (upper/lower MAC) and the various subprotocols such as 802.11i and 802.11e are the same, though of course a mirror image of the protocol functionality is implemented as compared to the client. However, there is also a large amount of additional firmware required for configuration, management, provisioning, recovery, and an interface to the user, either directly or through a WLAN switch. In some cases, quite a large amount of high-level protocol support (Telnet, DHCP, HTTP, RADIUS, etc.) is contained within the firmware image run by the AP.

A relatively recent trend in enterprises is the incorporation of multiple "virtual" APs within a single PHY AP. Essentially, each AP acts as several logical APs, broadcasting multiple beacons, advertising multiple service sets (with different SSIDs), and allowing clients to select a specific logical AP to which they would like to associate. The logical APs are frequently configured with different security settings, and virtual LAN (VLAN) facilities on the Ethernet side are used to direct traffic appropriately. The effect is to set up two or more "overlay" WLANs in the same area, without the expense of duplicating all the AP hardware; for example, an enterprise can deploy a guest network for use by visitors and a well-protected corporate network for use by its employees with the same set of APs.

With the spread of WLANs in consumer and multimedia applications, a number of special-purpose variants of APs have been developed. The most common one, of course, is the ubiquitous wireless gateway: a combination of AP, Ethernet switch, router, and firewall, normally used to support home Internet service. Other devices include ADSL and cable modems with the AP built into them (i.e., simply replacing the Ethernet spigot with an appropriate broadband interface), and wireless bridges or range extenders, that relay WLAN packets from one area to another. All of these devices use much the same structure as that of a standard AP, changing only the firmware and possibly adding a different wired interface.

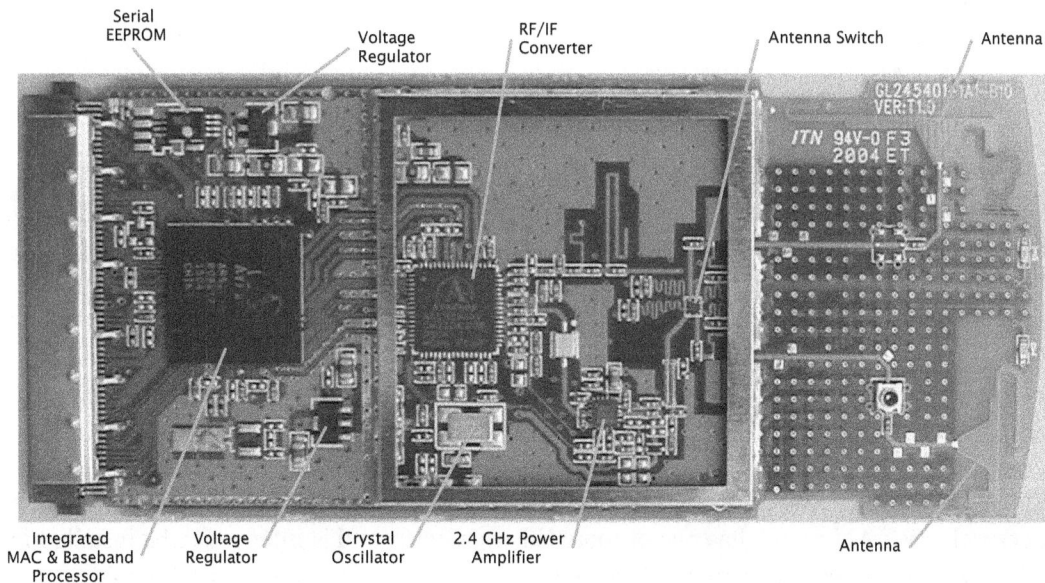

Figure 1.11: A WLAN NIC Chipset

1.3.3 WLAN Switches

Of interest for enterprise situations is the trend towards "thin APs". This basically means that a large fraction of the higher-layer 802.11 functions, such as connection setup and mobility, are centralized in a WLAN switch rather than being distributed over individual APs. (Some vendors refer to the WLAN switch as a "WLAN controller".) The CAPWAP protocol described previously is being standardized to enable the APs and WLAN switches to communicate with each other. From a hardware point of view a "thin AP" is not significantly different from a normal or "thick" AP, and in fact at least one vendor uses the same hardware for both applications, changing only the firmware load.

The benefits of "thin APs" and centralized management are not difficult to understand. When an enterprise deploys hundreds or thousands of APs, manual configuration of each AP becomes tedious and expensive, particularly considering that APs are often stuck in hard-to-reach or inaccessible places such as ceilings and support columns. The "thin AP"/ WLAN switch model, on the other hand, enables the enterprise network administrator to set up a single configuration at the switch, and "push" it out to all of the APs at the same time. Firmware upgrades of APs become similarly easy; once the WLAN switch has been provided with the new firmware, it takes over the process and "pushes" the firmware down to all the APs, and then manages the process of reloading the configuration and verifying that the upgrade went well.

The following figure shows a typical switch-based WLAN architecture.

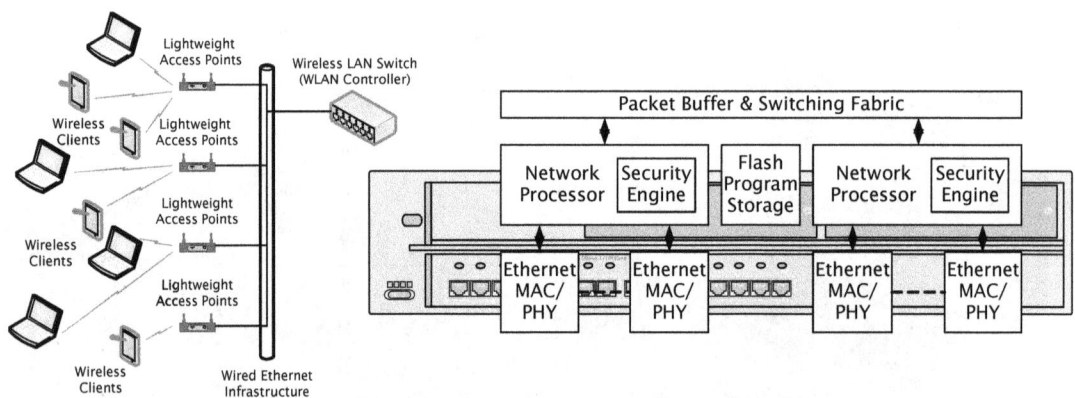

Figure 1.12: WLAN Switch Architecture

In general, a WLAN switch has one or more Ethernet ports, and is intended to be installed in a wiring closet or equipment center. APs may be connected directly to the switch ports, or (more commonly) to an Ethernet LAN infrastructure to which the WLAN switch is also connected.

For example, a hierarchy of LAN switches may be used to connect a large number of APs, up to a hundred or so, to a single port of a WLAN switch.

There is an emerging trend among large equipment vendors such as Cisco Systems to integrate the WLAN switch directly into a high-end rackmountable wiring closet or data center Ethernet switch. In this case, either a plug-in services card is provided with the WLAN switch hardware and firmware on it, or else a factory-installed plug-in module is used to support the WLAN switch hardware and firmware.

The protocol run between the WLAN switch and the AP tends to vary by vendor, with many custom extensions and special features for proprietary capabilities. As previously mentioned the CAPWAP group at IETF is standardizing this protocol. In all cases, however, the protocol provides for the following basic functions:

1. discovery of the WLAN switch by the APs, and discovery of the APs by the WLAN switch;

2. firmware download to the AP;

3. configuration download to the APs (e.g., SSIDs supported, power levels, etc.);

4. transport of client association and security information;

5. transport of client data, in cases where the data path as well as the control path passes through the WLAN switch.

1.4 The RF Layer

The RF layer of the WLAN protocol is, of course, the *raison d'etre* of every WLAN device; it is this layer that provides the "wireless" connectivity that makes the technology attractive. This section will briefly summarize the requirements placed on transmitters and receivers intended for WLAN service that go beyond standard radio transceiver needs. The reader is referred to one of the many excellent introductory books on the WLAN RF layer, such as *RF Engineering for Wireless Networks* by Dobkin, for further information.

1.4.1 Transmitter Requirements

Transmitters for typical 802.11 WLAN devices are required to produce 50 mW or more of power output in the 2.400–2.483 GHz and possibly also the 5.150–5.825 GHz frequency bands. The following figure shows the general frequency bands and emission limits in various countries.

The early 802.11 transmitters were relatively uncomplicated devices, as they were required to transmit BPSK or QPSK modulation at 1 or 2 Mb/s in a 16 MHz channel bandwidth – not very

	2.400–2.4835 GHz	4.900–5.091	5.091–5.150	5.150–5.250	5.250–5.350	5.470–5.725	5.725–5.825
USA	1000 mW		100 mW	40 mW	200 mW		800 mW
Europe	100 mW			200 mW		1000 mW	
Japan	50 mW	250 mW		200 mW			

GHz: 2.400 2.412   2.471 2.4835 2.497   4.900 4.940 4.990 5.091   5.150   5.250   5.350   5.470   5.725   5.825

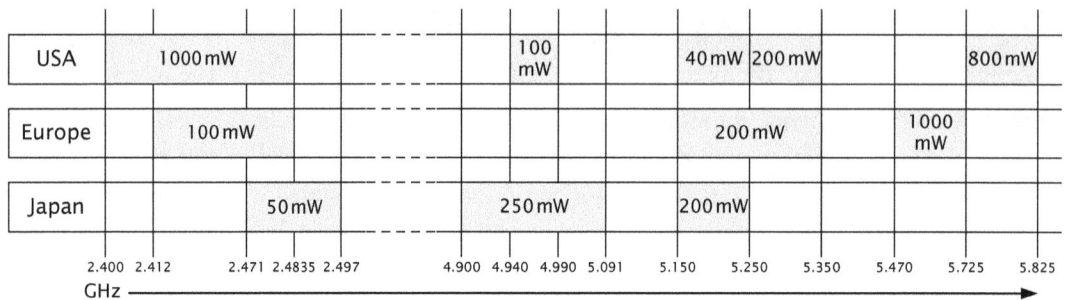

Figure 1.13: 802.11 Frequency Bands and Emission Limits

exacting requirements. The 802.11a and 802.11g standards, however, raised this to 54 Mb/s in the same bandwidth. In order to support these PHY rates in the typical indoor propagation environment, it was necessary to use complex modulations – 64-point QAM constellations – with OFDM. The design of an 802.11a or 802.11g transmitter is therefore far more complicated. (Of course, the design of a MIMO transmitter for the 802.11n draft standard is more complicated still.)

The key issue in supporting OFDM modulation is the high peak-to-average power ratio resulting from the modulation. A typical FM transmitter has a peak-to-average ratio of 1(0 dB); that is, the output is virtually a continuous sine wave. By comparison, an OFDM signal can have a peak-to-average ratio of as much as 8 dB. If the transmitter, particularly the power amplifier, is incapable of handling these peaks without clipping or compression, the resulting non-linear distortion will produce two adverse effects:

1. The output spectrum will widen due to the mixing and production of spurious signals.

2. A higher rate of bit errors will be generated at the receiver.

The spectral purity of 802.11 transmitters is strictly regulated (and specified in the 802.11 standard) in order to prevent adjacent channel interference. Spectral purity is represented by a spectral mask, which is simply the envelope in the frequency domain of the allowable signal components that can be transmitted.

One simple means of assuring a high-linearity transmitter is to ensure that the peak power output is always much less than the compression level of the power amplifier (PA) and driver chain. Unfortunately the peak-to-average ratios of OFDM means that obtaining a sufficiently high average output power requires a rather large and expensive PA. Designers therefore spend a great deal of time and energy attempting to strike a good balance between cost, size, and output power.

Beyond linearity, power consumption and cost are probably the most significant factors to be considered by 802.11 transmitter designers. All of the modulation functions are normally

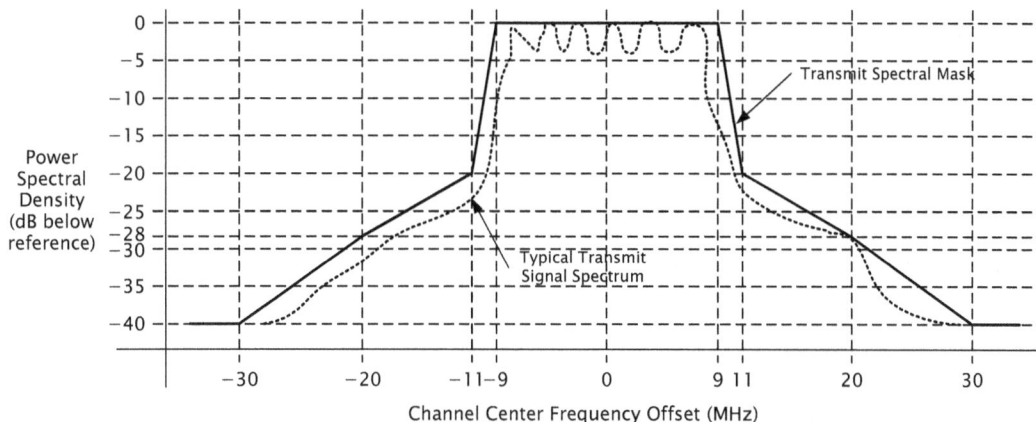

Figure 1.14: OFDM Transmitter Spectral Mask

carried out using digital signal processing at baseband, and the signals are then up-converted to the operating frequency band. The complex digital processing required by OFDM consumes both power and chip die area. Further, a high-output low-distortion PA chain consumes almost as much power as the rest of the radio combined. Minimizing power consumption is therefore high on the list of design tradeoffs. (It is noteworthy that one of the biggest impediments to the use of 802.11a and 802.11g technologies in VoIP-over-WLAN handsets is power consumption; the older 802.11b radios consume a fraction of the power of an 802.11g system.)

A key parameter that is a consequence of the TDD nature of 802.11 is the transmit-to-receive (and vice versa) switching delay. To maximize the utilization of the wireless medium, it is desirable for the interval between transmit and receive to be kept as short as possible: ideally, well under a microsecond. This in turn requires the transmitter in a WLAN device to be capable of being ramped from a quiescent state to full power in a few hundred nanoseconds, without burning up a lot of DC power in the quiescent state, which is not a trivial engineering challenge.

1.4.2 Receiver Requirements

The principal burden placed on an 802.11 receiver is the need to demodulate data at high rates (54 Mb/s) from a many different transmitters (thanks to the shared-medium channel) with a low bit error ratio.

The 802.11 PHY standards provide for special training sequences or preambles that precede every packet. The receiver must constantly scan for these training sequences, lock on to the (known) information within them, and use them to fine-tune the oscillators, A/D converters, and demodulator parameters. For example, 802.11 A/D converters have only 5–7 bits of resolution, to save power and cost; thus the receiver makes an accurate measurement of average power level during the training sequence, and uses this value to center the signal in the A/D converter's limited operating range.

Unlike their more complicated brethren in the cellular world, 802.11 devices do not make use of more advanced techniques such as Rake receivers and combining diversity. (This is changing with 802.11n, however.) The key engineering tradeoff in WLAN receivers, therefore, is cost and power consumption versus error-free reception.

1.4.3 Rate Adaptation

Rate adaptation is an interesting peculiarity of the 802.11 PHY layer. To put it simply, an 802.11 PHY – under control of the lower level of the MAC – selects the best rate for data transmission under the prevailing propagation and interference conditions. It is to facilitate rate adaptation that there are so many rates defined for an 802.11g or 802.11a PHY; specifically, 1, 2, 5.5, 6, 9, 11, 12, 18, 24, 36, 48, and 54 Mb/s). It thus provides a dynamic and automatic method of adjusting the PHY rate to match the channel conditions.

Rate adaptation is basically a tradeoff between raw bit-level throughput and frame error rate. A high PHY rate such as 54 Mb/s can transfer data more than twice as fast as a lower PHY rate such as 24 Mb/s, but also requires a much higher signal-to-noise ratio (SNR) to maintain the same frame error ratio. We are, after all, interested in transferring correct data, not merely squirting bits across! When the SNR drops due to increasing range or interference level, transmissions at 54 Mb/s experience higher levels of frame errors, which in turn require more retransmissions – thus dropping the net effective data transfer rate. At some point, it is actually more efficient to use a lower PHY rate that is less susceptible to frame errors at that SNR; the reduced bit rate is compensated for by the lower retransmission rate, because the frame errors decrease. The PHY therefore adjusts its bit rate downwards to keep efficiency high.

The specific algorithm used to determine the rate adaptation behavior of a WLAN device is not standardized, and is usually vendor-specific and proprietary. In general the rate adaptation process looks at two parameters: the signal strength of the packets received from the counterpart device (e.g., in the case of a client this would be the beacons and packets received from the AP) as well as the perceived frame error ratio at the far end. The perceived frame error ratio is deduced by looking for missing acknowledgement packets (ACKs) in response to transmitted data frames, because 802.11 does not provide for any explicit indication of frame error ratio between devices. A lower signal strength, particularly coupled with a higher far-end frame error ratio, indicates a need to drop the PHY rate in order to maintain efficient data transfer.

Note that some (misguided) device vendors actually implement a sort of "reverse rate adaptation" algorithm; they configure their device to transmit at the lowest possible PHY data rate at all times, until the traffic load increases and the device starts dropping packets, at which point the PHY bit rate is ratcheted up. This, of course, leads to a substantial drop in efficiency for the WLAN as a whole.

1.4.4 Coexistence

All wireless devices, whether a simple AM radio or an 802.11 OFDM link, are subject to coexistence issues. Coexistence in this context refers to interference to, or from, other licensed or unlicensed radio services. As the number of such radio services occupying the microwave bands (particularly above 1 GHz) is increasing at a rapid pace, coexistence has become a significant issue; in fact, the IEEE has recently formed a separate group (IEEE 802.19) to monitor the coexistence issues of all of the different types of wireless communication standards being created within the 802 committee.

The most notorious example of coexistence issues observed with WLANs is, of course, interference from microwave ovens. However, many other situations exist, particularly in the 2.4 GHz band which is shared by a large variety of users. For instance, Bluetooth devices also use the 2.4 GHz band; their frequency-hopping radios can sometimes shut down wireless links. 2.4 GHz cameras and video links, not to mention cordless phones, can affect (and be affected by) WLANs. In the 5 GHz band, particularly in Europe, WLANs are secondary to certain types of radars; as a consequence, 802.11a radios implement radar detection mechanisms to detect and avoid radar signals.

1.4.5 Propagation

Wireless links are extremely subject to propagation conditions between the transmitter and receiver. (Wired networks have the luxury of essentially ignoring this issue; if optical or twisted-pair cables are properly installed, then the user is assured of extremely high SNR on a permanent basis.) Indoor propagation at microwave frequencies is particularly influenced by all sorts of changes in the environment surrounding the wireless devices. It is not unusual for the propagation characteristics of an office environment to change drastically between daytime, when there are lots of occupants busy absorbing microwave energy, and nighttime, after everyone has gone home.

Propagation issues generally increase as the wavelength drops; thus 5 GHz WLANs have a comparatively lower range than 2.4 GHz WLANs, due to absorption in the walls and doors as well as the increased impact of diffraction and fading. Further, the multipath effects within buildings leads to inter-symbol interference (ISI) that limits the data rate possible over the wireless link: 802.11 WLANs deal with this issue at the higher data rates by resorting to OFDM modulation, which increases the symbol period (to $4\,\mu s$) to minimize ISI and adds in a guard interval between symbols to let multipath settle out.

A whole science has been built around the modeling of indoor propagation effects as well as the actual measurement of propagation characteristics of indoor environments and their impact on wireless communication channels. The reader is referred to the excellent books by Durgin (*Space-Time Wireless Channels*) and Rappaport (*Wireless Communications: Principles and Practice*) for more information on this subject.

1.4.6 Multiple Input Multiple Output

The upcoming 802.11n draft standard uses MIMO techniques to support nearly an order of magnitude increase in the PHY data rates of 802.11 links. Simply put, MIMO takes a disadvantage (multipath effects within buildings, caused by signals scattering off metallic objects) which reduces data rates in 802.11g or 802.11a, and actually converts it to an advantage by employing the multipath to increase data rates. There IS such a thing as a free lunch!

At the frequencies used in WLANs (2.4 GHz and up, with wavelengths of 12.5 cm or less), even small metallic objects can reflect or diffract (i.e., scatter) the energy propagating from the transmitter to the receiver. A typical indoor environment is thus full of scatterers of all kinds, which result in multipath propagation between transmitter and receiver, as shown in the following figure.

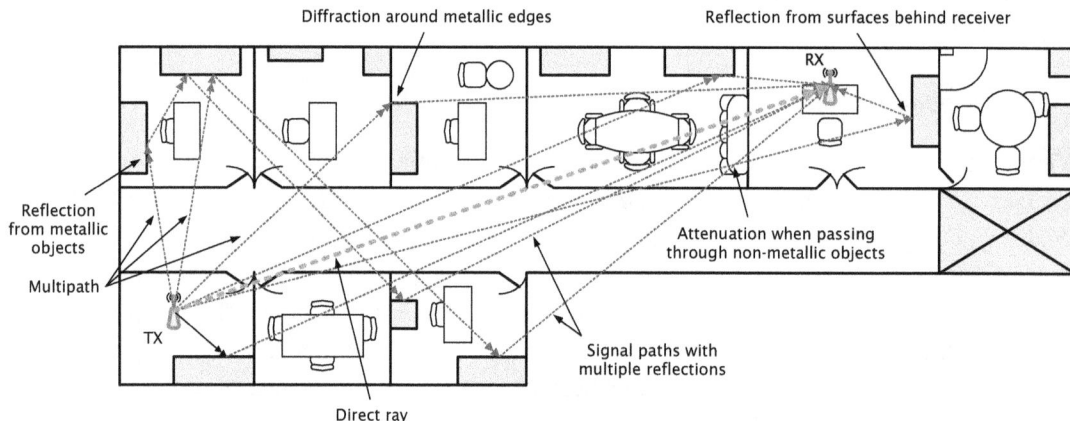

Figure 1.15: The Indoor Channel

Normally, this multipath is a nuisance; energy arriving over different paths may be just as likely to cancel each other (destructive interference) as to reinforce each other (constructive interference), leading to fading effects and frequency-selective channels, all of which limit the range and data rates of conventional receivers and transmitters. However, it was observed in the late 1960s that the multiple signal paths could actually be used to increase the bandwidth provided that they were uncorrelated, that is, the amplitude and phase of the different multipath signals are statistically independent. In essence, one can regard the multiple signal paths as being multiple independent parallel radio channels, and send different signals down these channels; the effect is that the available bandwidth is increased by the number of such radio channels, even though all of these channels are in the same frequency band and the same physical space. This is the basis for the MIMO technique.

A simplified view of the MIMO process is as follows: take the source data signal, split it up into as many smaller pieces as there are uncorrelated signal paths, and transmit each piece down a separate signal path. At the receiving end, all of the individual pieces are received and then reassembled into the original data signal. Effectively therefore, the bandwidth of the channel has increased by N, where N is the number of signal paths. (This is also the basis for the term MIMO – the radio channel is regarded as having multiple inputs and generating multiple outputs.) This is represented graphically in the figure below.

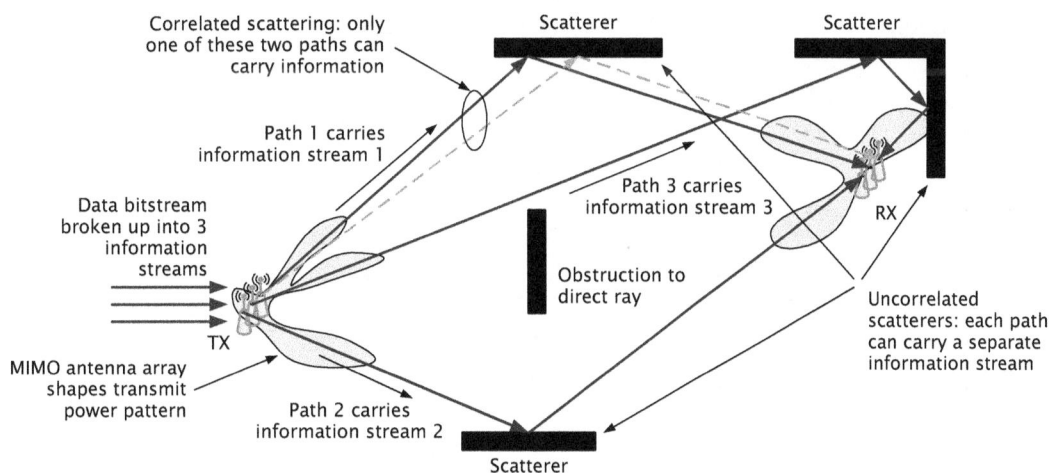

Figure 1.16: Using Uncorrelated Multipath

It should be kept in mind that this is actually a rather rough approximation to the real way in which the MIMO process is performed – the transmitter does not locate individual scatterers and shoot beams off each one. However, it is sufficient for an understanding of the basis of the process.

In order to send different pieces of information down the different signal paths, both the transmitter and the receiver must be able to distinguish between the various paths. This is done by equipping the transmitter and receiver with multiple antennas, each connected to a completely separate but synchronized radio. In the case of the receiver, signals arriving from different directions can be distinguished (decorrelated) by using the small differences in phase and amplitude of the same signal received by different antennas separated in space. A reverse process is used at the transmitter, where the separate streams are fed to the set of transmit antennas with slightly different phases and amplitudes, so that they essentially use different paths through the channel to arrive at the receiver. The combination of these two effects causes a system with N transmit antennas and N receive antennas to have N times the bandwidth of a single-antenna system. Further, the MIMO technique also offers a considerable reduction in interference, as external interferers will not decorrelate in the same way as the desired signal and hence will be largely filtered out.

There is, of course, a great deal of additional complexity involved with a MIMO system; for example, the receiver and transmitter have to co-operate to measure channel properties, before transmitting or receiving data. The 802.11n draft standard therefore includes a good deal of complexity around the channel estimation and MIMO signal processing functions. However, the prospect of an order of magnitude improvement in throughput in the same spectrum (and with the same transmitted power levels) is expected to make WLANs attractive for many indoor situations where only fixed LAN connections are usable today.

Metrology, Test Instruments, and Processes

This chapter serves as a brief, and hopefully painless, introduction to the basic concepts and terminology of metrology, with emphasis on those topics that are most pertinent to wireless LAN (WLAN) devices. The various types of test instruments generally used during all phases of the design and testing of WLAN devices, systems and networks are also presented, with some indication as to what each piece of test equipment might actually be used for.

2.1 Metrology: the Science of Measurement

Metrology is defined as "the science of measurement". The term is generally applied to both experimental and theoretical determinations of values, usually within scientific and technical fields. The science of metrology predates the Industrial Revolution; indeed, it was essential for the mass-production methods that led to the modern manufacturing.

2.1.1 Quantitative vs. Qualitative

The terms "quantitative" and "qualitative" both refer to means of analyzing or describing the properties of something, typically a physical or logical entity. In qualitative analysis, the properties are described in terms of classifications or attributes, or assessed relative to something else. For example "color" is a qualitative property, and is expressed in terms of arbitrarily assigned tags such as "red", "orange", and so on. A quantitative property, however, exists in a range of magnitudes, and can therefore be measured and expressed in numeric forms. Continuing the example, a quantitative version of "color" (for physicists and photographers) is color temperature, which is the temperature in Kelvin to which a black body must be raised to emit light of the equivalent wavelength.

Engineering measurements normally focus on the quantitative rather than the qualitative attributes of devices and systems. (Technical marketing assessments are a notable exception!) Quantitative measurements avoid subjective differences; if performed properly, they yield the same result regardless of who is making the measurements. Further, quantitative measurements are reproducible and transportable; measurement setups can be created in different places to measure the same parameters, and can be assured of getting the same results.

2.1.2 Direct vs. Indirect

Metrology encompasses both direct and indirect measurements. A direct measurement is one which is measured with an instrument that can directly convert the quantity of interest into human-readable form; for example, a DC voltage measurement is usually a direct one. An indirect measurement, however, involves inferring or calculating the measured quantity from one or more direct measurements. DC power, for instance, is often measured indirectly, by measuring the voltage and the current and then multiplying the two. An indirect measurement on a device also involves creating a model of how the device functions, in order to relate the direct measurements to the desired indirect value.

In many cases both direct and indirect measurements can be used to obtain the same quantity. Direct measurements are preferred over indirect measurements, as they reduce the possible sources of error and avoid the stacking of measurement uncertainties.

2.2 The Nomenclature of Measurement

The science of test and measurement, in keeping with every other engineering and scientific discipline, has produced its own nomenclature and jargon. While a complete glossary of terms is out of the scope of this book, some of the more important terms are defined and explained here as an aid to understanding. Those who would like to delve further into such details are encouraged to start with the National Institute of Standards and Technology (NIST) Technical Note 1297, "Guidelines for Evaluating and Expressing the Uncertainty of NIST Measurement Results", as well as numerous well-written and detailed whitepapers on the subject published by Agilent Technologies on its Website.

2.2.1 Device Under Test

The device under test, normally abbreviated as DUT, is the entity being tested when making a measurement. This is the preferred nomenclature when a single component, device, or system is being tested.

2.2.2 System Under Test

A system under test (SUT) refers to a group of entities being tested as a single system. This is the preferred nomenclature when several distinct devices or systems are interconnected and tested as a unit. See Figure 2.1.

2.2.3 Calibration and Traceability

Every reputable measuring instrument undergoes a process of calibration, which usually attempts to relate the measured values to standard values for the same parameter as established by a national or international standards body. For instance, consider the

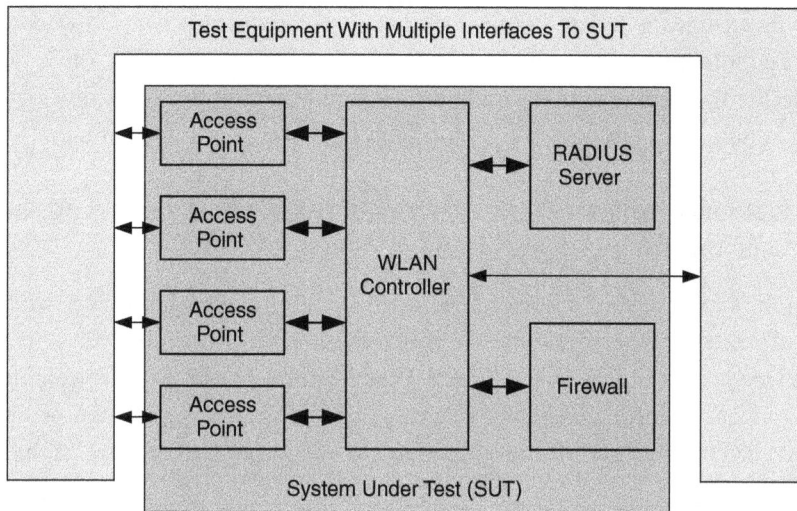

Figure 2.1: Example of a System Under Test (SUT)

measurement of voltage. The primary definition of the volt is maintained as a Josephson voltage standard at various national standards bodies, including the NIST. A high-precision voltmeter would then be calibrated against the voltage standard; as it derives its measurements from the voltage standard, its calibration is referred to as being traceable to the standard ("NIST-traceable calibration").

Primary standards (such as the NIST-maintained Josephson voltage standard referred to above) are the "golden references", forming the basis for all quantitative engineering measurements performed worldwide. They are maintained with great care, and access to them is quite difficult. Working standards are set up by people interested in actually calibrating instruments, such as test equipment vendors. Transfer standards form the bridge between the two; they are calibrated against the primary standards, and then used to ensure that working standards are correctly calibrated.

Note that another instance of the term "calibration" is commonly used in the RF industry. This refers to the process of factoring in ("calibrating out") the effects of the test fixtures, connectors, and cables before making a measurement on a DUT. For example, when performing an impedance measurement using a microwave network analyzer, a process of fixture calibration is performed to ensure that the final measurement will reflect the properties of the DUT rather than those of the test setup.

2.2.4 Conformance, Performance, and Interoperability

The terms conformance, performance, and interoperability are often confused in the networking industry, so a short explanation is in order.

Conformance determines whether a device adheres to the requirements of a standard or specification. A conformance test yields only a "true" or "false" ("yes" or "no") answer, indicating whether the device adheres to a particular requirement or not. Conformance to a standard rarely guarantees any level of performance; standards are written mainly to allow manufacturers to implement devices that can interface to each other (interoperate). Further, conformance is absolute; a device either conforms or it does not, it cannot be said to "almost conform" or "conform well".

Performance refers to how well a device functions, usually relative to other devices. In direct contrast to conformance, a performance measurement has no "right" answer, only "better" answers. A device can exhibit high performance without being either conformant to a standard or being able to work (interoperate) with other devices; in fact, it is not unknown for vendors to deliberately introduce non-conforming features to boost the performance of their products.

Interoperability assesses how well a device from one vendor works in conjunction with a related device from another vendor. For example, it is necessary for an access point (AP) and a client to interoperate in order to transfer data over a WLAN. An interoperability test only indicates whether a combination of devices works well as a unit; it does not say anything about the quality of each individual device. Further, an interoperability test performed on one combination of devices says nothing about how well some other combination of devices will function; it is necessary to test every combination separately.

Most end-users are primarily interested in performance and interoperability. (As marketing folks well know, performance sells.) Conformance is only interesting insofar as the lack thereof interferes with performance and interoperability.

2.2.5 Functional vs. Regression Testing

A functional test is aimed at testing specific aspects of a device or system. For example, a test to verify that an AP can forward 802.11 packets correctly would be a functional test. Such tests are typically conducted when a device is first developed, and are frequently performed manually.

A regression test is normally a set of functional tests that are run repeatedly as devices change (e.g., due to new firmware releases) to ensure that existing functionality has not been broken by the changes. Typically, once device functionality has been proven by functional testing, the same tests are incorporated into a regression test suite. Regression tests are frequently automated, sometimes with elaborate software frameworks that are the subject of considerable development effort by vendors of WLAN devices.

2.2.6 Benchmarking

Benchmarking is the process of conducting a specific set of performance tests on a device or system to assess its suitability for a particular user application. The results of the performance tests ostensibly translate to the perceived user experience of an end-user of the product. Benchmarking is often carried out in a comparative manner, to rank several products relative to each other using quantitative measures. Benchmarking may also be used by vendors to claim that their products are better than others ("benchmarketing"), and by trade journals to provide equipment selection recommendations to their readers. A variant of benchmarking, also known as equipment qualification, is conducted by large customers of networking equipment to verify vendor performance claims prior to purchase decisions.

Good benchmarking tests try to approximate real-usage scenarios and user environments as closely as possible, so that the results can be used by end-users to guess at the performance they can expect. Obviously it is not possible to obtain the exact user setup, but careful attention to the test configuration can come quite close.

Benchmarking is almost exclusively the domain of performance testing; conformance and interoperability tests are very rarely used.

2.3 Measurement Quality Factors

The "quality" of a test or measurement is determined by how closely and consistently it comes close to the actual value of the property being measured. Terms such as "uncertainty", "accuracy", and "precision" are all used and misused in connection with metrology; this section will define these terms.

2.3.1 Uncertainty

Uncertainty is a key parameter in the quality of a measurement. Formally, uncertainty is defined by the ISO somewhat obscurely as "a parameter associated with the result of a measurement that characterizes the dispersion of values that could reasonably be attributed to the measurand". Basically, uncertainty is an estimate of the maximum possible deviation of the measured results from the actual value; it sets the bounds between which the "true" value lies.

Uncertainty is not the same as error. An error by definition is known, and hence can be compensated for and removed; uncertainty cannot be removed, except by improving the measuring apparatus.

2.3.2 Accuracy

Accuracy refers to the degree of agreement between the result of a measurement and the actual value of what is being measured. Modern metrology tends to focus more on uncertainty than accuracy, because uncertainty encompasses not only accuracy but also precision.

2.3.3 Resolution

The resolution of a measuring instrument is the minimum difference in readings that two different measured values can produce. As an example, a yardstick might be marked off in tenths of an inch; the resolution of the yardstick is therefore 1/10 in. Note that resolution does not imply either accuracy or precision – the yardstick might be only accurate to an inch, even though it has markings of 1/10 in.

2.3.4 Precision, Repeatability, and Reproducibility

Precision is the degree to which the results of two or more measurements of the same quantity will agree with each other, and is usually characterized in terms of the standard deviation of a number of measurements. Note that nothing is implied about the accuracy of the results. A measurement can be precise without being accurate. To take an example, a digital voltmeter used to measure a 1.5 V DC voltage might indicate successive readings of 1.700, 1.698, and 1.701 V; the results are thus quite precise, but not very accurate.

Precision is usually separated into two components: repeatability and reproducibility. Repeatability indicates how well the results of successive measurements of the same quantity agree when made under the same conditions of measurement. Reproducibility refers to the agreement of measured results with each other under different conditions of measurement (e.g., performed with a different test instrument at a different location and by a different operator).

The following figure uses the familiar "arrow and target" analogy to illustrate uncertainty, precision, and resolution.

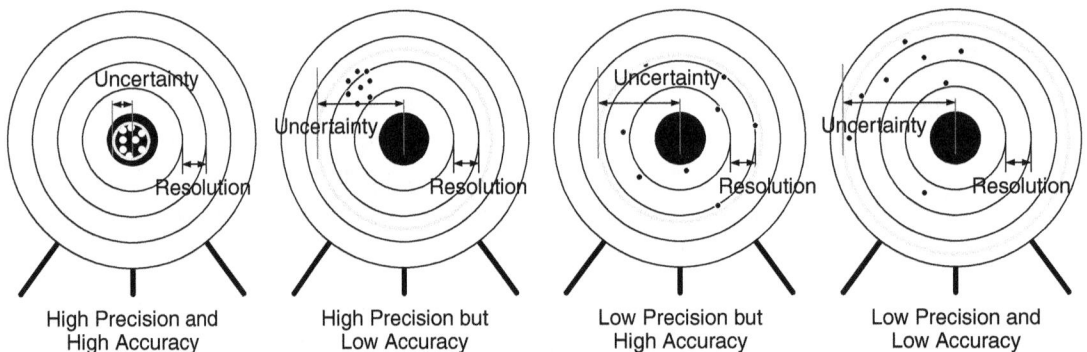

Figure 2.2: Uncertainty, Precision, and Resolution

2.3.5 Validity

A valid test is one that measures what it is supposed to measure. Validity implies repeatability and consistency, but the converse is not true; a test can be precise and repeatable, but completely invalid. For example, a voltmeter with a low internal resistance, when used to measure the DC voltage in a circuit, can draw so much current that the behavior of the circuit actually changes; in this case, the voltage measurements are invalid, even though they may be quite repeatable and even accurate.

2.4 The WLAN Engineer's Toolbox

Engineers who develop, verify, or install WLAN devices and systems have a huge variety of test equipment at their disposal. However, the plethora of equipment can be confusing to someone who is new to WLANs, or who is only conversant in a subset of the WLAN space. This section surveys the space of WLAN test equipment and briefly mentions the purpose of each type of equipment. (Note that basic tools that every engineer is expected to be familiar with, on the level of oscilloscopes and multimeters, are not included.) The reader is referred to the appropriate equipment manuals, or books specifically on RF test equipment, for more details on what each piece of gear does and how it is used.

2.4.1 Physical (PHY) Layer Test Equipment

The RF PHY layer is necessarily a complex field and consequently a large variety of equally complex test equipments have been developed by vendors such as Agilent, Tektronix, and Anritsu to help engineers deal with RF and digital issues.

Signal generators are used to produce accurate RF test stimuli (with various types of modulating signals imposed) at different frequencies, during equipment development and verification. Vector signal generators (VSGs) are a recent development, comprising signal generators supporting the complex modulation schemes used in WLAN communications, even down to generating streams of fully formed data packets. The opposite side of the coin comprises the spectrum analyzer (and its close cousin the Fast Fourier Transform (FFT) analyzer) that allows complex RF signals to be decomposed into their spectral components. The vector signal analyzer (VSA) is a special type of spectrum analyzer that actually demodulates injected WLAN signals, making measurements such as occupied bandwidth and error vector magnitude (EVM).

RF power meters are used to make peak and average power measurements on transmitters, and to calibrate receivers. Note that a spectrum analyzer can also be used to make such measurements, but a power meter is generally cheaper and more accurate. A related instrument is a directional wattmeter, which can make both power and antenna mismatch (SWR) measurements. Sometimes the functions of VSG, VSA, frequency counter, power meter, SWR meter, etc. are combined into one unit, referred to as a communication test set or communication signal analyzer. Field strength meters are used during antenna pattern measurements and electromagnetic compatibility (EMC) testing; they measure the radiated field strength around a DUT.

A network analyzer is customarily used during device design and development to characterize the frequency response (transmission and reflection) properties of a device or network. Network analyzers are also useful for characterizing environments during propagation studies; one means of obtaining the frequency response of an RF channel is to use a network analyzer with a pair of calibrated antennas. A time-domain reflectometer (TDR) performs a similar function to a network analyzer, but produces the impulse response of the device in the time domain, and is generally used for testing cables.

On the digital side of life, there is not quite such a variety of tools. The most commonly used tool is the logic analyzer, which allows design engineers to look capture signals in parallel from many different points in a digital system. A number of special-purpose tools have also been developed around the standardized buses and embedded CPUs that are normally used; these are referred to as emulators (bus emulators when applied to standard interface buses) and bus analyzers.

Besides the plethora of hardware tools, a large number of software tools are also available. These are used primarily during design and development, as (verification and Quality Assurance (QA) people prefer to focus on the physical DUT rather than a simulation thereof). An exhaustive description of the variety of software tools is outside the scope of this book, but the main types of tools involve the following:

- General algebraic/matrix math packages such as Matlab. These are normally used to simulate and develop digital signal processing algorithms.

- Circuit simulation packages such as Spice, which enable analog circuits to be soft-prototyped and tested.

- Microwave circuit analysis packages such as Sonnet. These enable microwave design engineers to design high-frequency circuits, including stripline and microstrip filters, amplifiers, etc. Related packages perform finite element modeling (FEM) functions, allowing the design of antennas and shielding.

Many other specialized programs are used as well, for PLL, amplifier, oscillator, and filter design and characterization.

2.4.2 Protocol Test Equipment

After the RF engineer has finished with a WLAN device, the digital hardware and software/ firmware engineers enter the picture. At this point, the test equipment needs to change from injecting and analyzing accurate RF signals to dealing with frame data, and even with entire high-level transactions. Protocol test equipment fills the bill here.

In the wired world, the "Swiss Army Knife" of protocol testing are the traffic generators/ analyzers (TGA) made by companies such as Agilent, Spirent, and Ixia. This is becoming true of the wireless world as well, with products from vendors such as VeriWave, but this is a more recent development. A TGA enables the development and validation/QA engineers to generate stimuli in terms of streams of frames with adjustable characteristics (length, rate, contents, etc.) and to measure and interpret the response of the DUT as streams of frames as well. Generally, the frame streams can be made quite complicated, in order to emulate the actual behavior of the devices and systems with which the WLAN DUT will be expected to interwork. In most cases, protocol testers must be sophisticated enough to implement the handshakes and exchanges of real devices, while supporting the analysis capabilities at multiple protocol layers that are necessary to untangle all this complexity for the benefit of the engineers using them.

An interesting middle-ground in the cellular area is the wireless protocol test set, which combines some features of low-level RF instrumentation (e.g., accurate power measurements) with many of the features of a full-blown traffic generator. These test sets are principally aimed at testing devices at Layer 2 or below; in many cases they enable the test engineer to "script up" precise sequences of frames to perform low-level handshakes. However, they are usually too low-level for much more than Layer 2 testing.

Protocol test equipment is not restricted to hardware-only devices; at the expense of accuracy and capacity, software programs running on standard PCs or workstations can be employed to perform similar tasks. Software traffic generators (e.g., Chariot and Qcheck from Ixia, or the Iperf freeware tool) are available for basic traffic generation and analysis tasks, and are especially useful for gaining some insight into the performance of an installed network. Software protocol analyzers (also known as "sniffers", from the eponymous Data General "Sniffer" product) are also widely used for capturing and analyzing frames from live networks; vendors such as AirMagnet or WildPackets provide these, and there are a variety of open-source or freeware versions as well. Most of these products, however, make no guarantees about keeping up with the full network capacity; it is not unusual to find that a software sniffer has missed a small percentage of frames while attempting to capture traffic, especially at high loads.

Specialized protocol testers and analyzers are also used for application layer test work; for instance, testing and verifying Session Initiation Protocol (SIP) stacks for voice applications, analyzing wireless security implementations, and so on. These also tend to be implemented as software on standard computers.

2.4.3 Installation Test Equipment

Pre- and post-installation test of WLANs is a significant part of the overall scope of WLAN testing. Unlike wired LANs, which does not require much more than an optical or copper TDR analyzer during deployment, WLANs need fairly extensive planning and checking in order to ensure a successful install. In addition to simple coverage analysis, WLAN installers need to manage interference, congestion, coexistence with other spectrum users and frequency (channel) planning. Equipment is generally available, but the WLAN installation process is still evolving and today is largely ad hoc.

The most common test tool used by thousands of small installers worldwide is simply a laptop with a WLAN card plus an AP. (Most WLAN cards come with vendor software that makes approximate measurements of received signal strength and in-band noise.) However, more sophisticated installers have recourse to portable spectrum analyzers specifically designed for the Industrial, Scientific, and Medical (ISM) and Unlicensed National Information Infrastructure (U-NII) bands, and field strength meters for general assessment of signal and interference levels. In some cases, stripped-down versions of communication test sets (normally combining a spectrum analyzer, a protocol analyzer, and a signal source) are used.

Vendors selling into the cellular industry supply "drive test systems", which are integrated test setups combining a variety of tools and software and designed for use in vans; the cellular operator then drives around a region before, during and after installation in order to assess coverage areas and perform capacity planning measurements. Equivalent test setups in the WLAN world are not (yet) commercially available, but some installation contractors have been known to create similar (but smaller-scale) ensembles on an *ad hoc* basis. These "walk-test" systems combine spectrum analyzers, sniffers, signal strength meters, global positioning system (GPS) systems, and even traffic sources to automate some of the installation tasks.

In addition to hardware test tools, some software tools are also available for assisting in the process of installation. Most large WLAN infrastructure vendors (e.g., Cisco Systems, Aruba Networks) offer proprietary packages that assist with the task of doing a site survey, which is an RF assessment of propagation and interference. Third-party vendors such as AirMagnet also offer site survey tools. Channel sounders for direct RF propagation measurements are usually beyond the capabilities of installers, but companies such as Wireless Valley and Motorola provide site modeling software, which if fed with a floorplan and descriptions will perform a software simulation to assess coverage and aid in siting APs. Of course, software sniffers are extensively used to locate neighboring WLAN systems and assess capacity of installed WLANs.

2.4.4 Other Test Equipment

In complete contrast to wired LAN/WAN developers, who rarely require anything beyond RJ-45 or optical patch cables in addition to their test equipment, WLAN testing involves the use of a bewildering variety of additional passive components.

RF cables – usually connectorized with SMA connectors – are obviously essential for any kind of conducted testing, along with Ethernet patch cables for APs. The most common item beyond this is a set of connectorized attenuators (usually fixed, though variable attenuators are also available) to equalize signal levels and avoid overloading receiver front ends. Power dividers/combiners are needed to connect, say, a spectrum analyzer to a DUT in addition to a traffic generator. Directional couplers are also used when power measurements need to be performed while testing; the directional coupler taps off a small, known fraction of the power traveling in a given direction so that a power measurement can be made without interrupting operation of the rest of the test setup. Dummy loads and terminators are frequently needed for terminating unused ports of dividers and DUTs; without proper termination, significant errors due to unwanted reflections can occur. Of course, due to the variety of RF connectors (SMA, TNC, N, BNC, and their reverse equivalents), a set of adapters is essential.

For open-air (or over-the-air testing within chambers) calibrated antennas and probes are necessary, in addition to the antennas that are provided with the DUT. Shielded testing necessitates RF enclosures, RF chambers, or even entire screened rooms. Microwave device testing with network analyzers requires test jigs and calibration standards. Finally, environmental (temperature and humidity) testing uses special chambers equipped with temperature and humidity controls in addition to test leads allowing the DUT to be energized and traffic to be passed.

To complete the contrast between wireless and all other forms of LAN testing, there is usually to be found in every test lab (sometimes with some preliminary hunting, unfortunately!) a torque wrench – for properly tightening the connectors on SMA cables. No such animal is required by any other networking technology.

2.5 Test Setups and Test Processes

A properly configured test setup, with all the required pieces (and none of the unnecessary ones) is essential for efficient and accurate tests on WLAN devices. The nature of the test setup obviously varies considerably with the nature of the test being performed, which in turn changes as the WLAN system progresses through the stages from design to installation. A few examples of test setups of wide applicability are introduced in this section; future chapters will devote much more detail to explaining these and other setups. The examples are provided in the rough order of their use in the actual WLAN design and verification process. We will begin by taking a brief look at how a WLAN device makes its way from a concept in an engineer's mind to a component of a fully functioning LAN.

2.5.1 WLAN Testing: Design to Installation

The process of developing and installing a WLAN is a long and complex one, with many stages along the way where extensive tests need to be performed. It also includes several different vendors who are likely to be in entirely different countries.

A full description of the process is outside the scope of this book, but a brief summary is very useful in relating the concepts discussed here to their actual application in the WLAN industry. The following figure shows the process.

a. *Chip design*: Every WLAN device (AP, client, etc.) contains a chipset of some kind. Companies such as Atheros, Broadcom, and Marvell design and manufacture such chipsets. Note that all WLAN chipsets in current use employ considerable amount of embedded firmware, which is usually developed in parallel with the silicon.

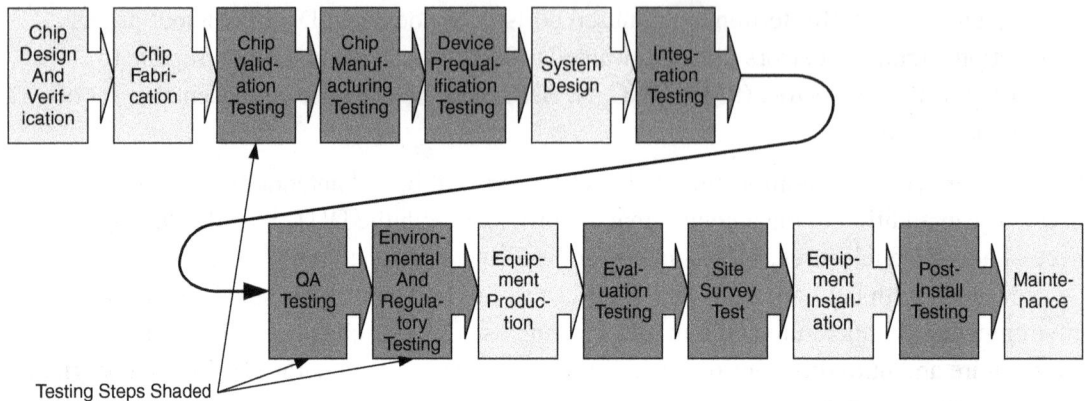

Figure 2.3: The WLAN Lifecycle

b. *Chip design verification*: The cost and time required to produce a single revision of a single chip being what they are, the chip design process is accompanied by a great deal of verification using simulation tests run on computers.

c. *Chip fabrication*: Once the design is complete, the chips in the chipset are manufactured at a fabrication plant ("fab") that processes silicon wafers into actual devices.

d. *Chip validation*: Simulation can cover only a very small fraction of the operational and usage characteristics of a chip, and thus the prototypes from the fab are subjected to an extensive testing process where they are placed on circuit boards ("reference designs") and run through a battery of tests. Bugs found during the validation tests frequently result in chip hardware revisions ("spins"), though in many cases firmware patches may be substituted. As the chip validation tests represent a microcosm of the system validation tests, many of the test setups and procedures described in this book apply.

e. *Chip manufacturing test*: Once a chip is committed to production, specialized and expensive high-volume test equipment is used to weed out defective parts during manufacturing. As the tests performed here are likewise specialized, they will not be considered in this book.

f. *Device pre-qualification test*: A modern networking system embodies a very large part of its functionality in the chipset; hence network equipment vendors frequently perform pre-qualification testing of devices from different vendors prior to designing them into actual equipment. A pre-qualification test normally consists of taking the reference design from the chipset vendor and running a series of performance or interoperability tests on it, using many of the test setups in this book.

g. *System design*: Once the chips to be used have been created and selected, the design of a complete piece of equipment (such as a client Network Interface Card (NIC), an AP, or a WLAN switch) begins. The system design usually integrates chips from different vendors and sometimes even the equipment vendor's own chips ("ASICs") for proprietary functions. The system design includes a considerable amount of software creation, along with tests on specific functions along the way.

h. *Integration testing*: After some months of system design and prototyping, the software and hardware pieces come together into a complete piece of equipment. Integration testing begins at this time, so that all of the components can be exercised as a unit. Integration tests are subsets of the full-scale performance and functionality tests that are carried out by the equipment vendor's QA lab, targeted towards known problem areas or complex functions.

i. *Verification and QA testing*: Once the complete system is functioning properly, but before release to production, the QA department of the system vendor must put the equipment through an extensive battery of tests of all kinds, ranging from simple performance measurements to extremely complex stress tests. QA testing usually follows a formal test plan and can take months to complete. Many of the test setups and test procedures covered in this book (apart from manufacturing and installation testing) will be applicable to QA test plans.

j. *Environmental and regulatory testing*: This phase focuses on the physical aspects of the equipment requirements: temperature, humidity, vibration tolerance, EMC, isolation, shock, lifetime and reliability tests. This phase of testing is normally carried out when the equipment is working well and nearing production.

k. *Equipment production*: Once all the testing has been carried out successfully, the equipment design is sent down to the manufacturing line, and production begins. Manufacturing test is performed on every copy of the produced equipment to ensure that it meets the design requirements; failing units may be fixed ("rework") or scrapped. Manufacturing tests use a small subset of the test procedures and equipment described in this book.

l. *Evaluation testing*: Large users of WLAN equipment, such as service providers and large enterprise IT departments, frequently conduct their own tests on WLAN equipment before actually deploying them in live networks. These tests are intended both to verify the manufacturer's claims of performance and functionality, and also to see how well the equipment performs when dealing with the user's own environment and applications. Evaluation testing is principally focused on comparative performance benchmarking, though some interoperability testing is also done, as WLAN equipment still suffers from compatibility issues.

m. *Installation*: Once the WLAN equipment has been evaluated and selected, it must be installed and configured, by either the enterprise IT staff or by installation contractors. A good installation process is preceded by detailed site and network planning, including an assessment of the user load on the network and the requirements of the specific applications to be supported. Once a plan has been drawn up, site survey testing is carried out to determine the sources of interference and locate APs to provide the necessary coverage. The WLAN infrastructure equipment (both wireless and wired) is put in place and wired up, after which there is usually a period of post-install testing with test traffic prior to allowing the network to carry live traffic.

n. *Maintenance*: All enterprise networks require ongoing maintenance and upgrade, as requirements change, users come and go, and equipment needs to be updated. This is usually the domain of the enterprise IT staff, who use software tools to monitor the health of the infrastructure and clients. In some cases – particularly in Internet service provider networks – active monitoring is done, where TGAs are used to run periodic performance and functionality checks on the running infrastructure, using some of the protocol tests described in this book.

It should be clear from the above that testing is done at nearly every stage of the lifecycle of WLAN equipment. Some of the categories of test procedures and test setups are described below.

2.5.2 Circuit Characterization

WLAN circuit characterization typically refers to the debug and alignment of the "front-end" RF/IF portion of a WLAN system, such as a chipset, an AP or a client card, prior to sending or receiving test signals. The most common setup used in such a situation is based on a microwave vector network analyzer, and can provide a great deal of information on the characteristics of the amplifiers, filters and mixers in the RF and IF signal paths. A typical network analyzer setup is shown Figure 2.4.

The setup is physically relatively simple, comprising a network analyzer, cabling, and special test probes to interface to the portions of the circuit under test. Prototype-boards may be assembled with special probe points, or containing only a subset of the components, in order to simplify probing. Critical aspects of the above setup are the design and use of the test probes and the calibration of the setup. Microwave RF engineers spend a lot of time and money on both of these issues.

2.5.3 Transmitter Measurements

Once the chipset or circuit has been characterized and is working properly, the next step is usually to send test data through the transmit paths and ensure that the transmitter (baseband and RF/IF) is functioning correctly. Generation of the test data can be done by either using the WLAN system itself to generate frames, or by configuring some sort of bit pattern generator for test purposes. If the WLAN system is used to perform the test data generation, then special software is usually created to send known sequences of data through the transmitter to simplify the task of the measurement equipment.

Figure 2.5 gives an example of a transmitter test setup, in this case for making EVM measurements.

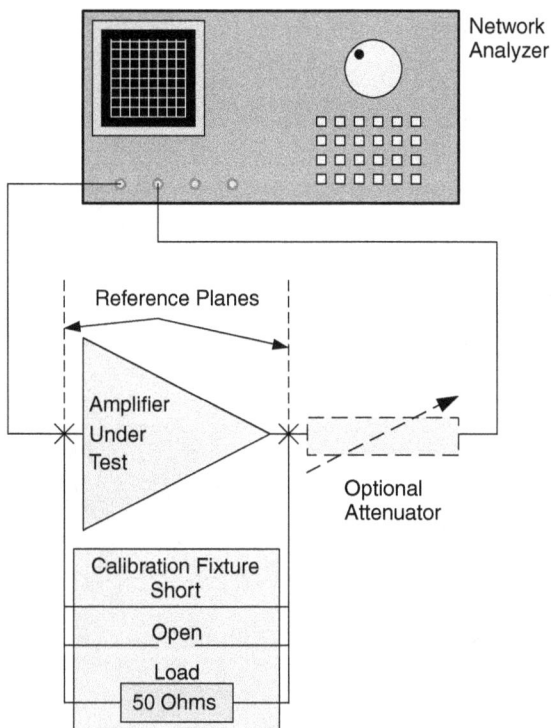

Figure 2.4: Typical Network Analyzer Setup

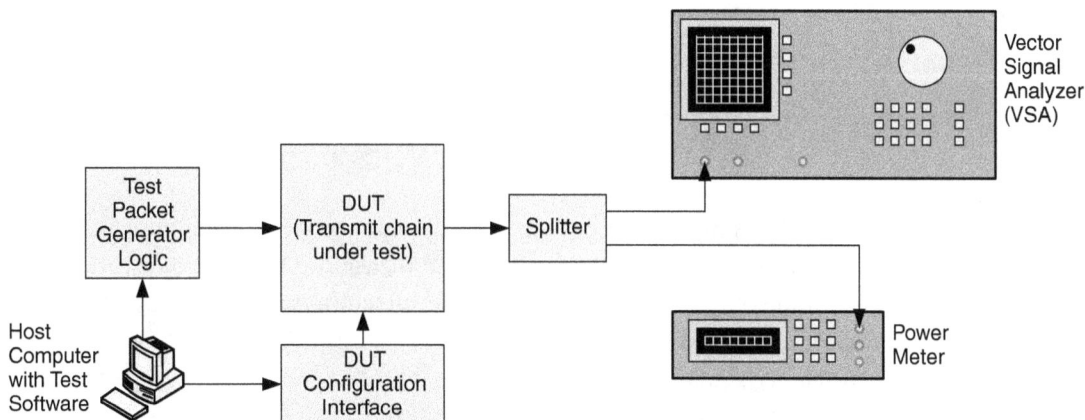

Figure 2.5: Example Transmitter Test Setup

As shown above, a typical EVM measurement involves primarily a spectrum analyzer capable of demodulating and analyzing WLAN signals (i.e., a VSA). The VSA locks on to the transmitted frame, decodes it, analyzes the signal, and displays a variety of useful information such as the EVM in percent, the frequency error, the average power level, and so on. Modern

VSAs are capable of a vast number of very useful measurements, and are indispensable in WLAN system RF test and alignment.

2.5.4 Receiver Measurements

As might be expected, characterizing a receiver requires the converse of transmit path characterization: modulated RF signals of known composition are injected into the antenna connector, and the demodulated digital output is sampled and checked. A typical test is to verify that the error rate (specifically, the frame error ratio or FER) meets maximum specifications at various input signal levels, and the setup for this is shown below.

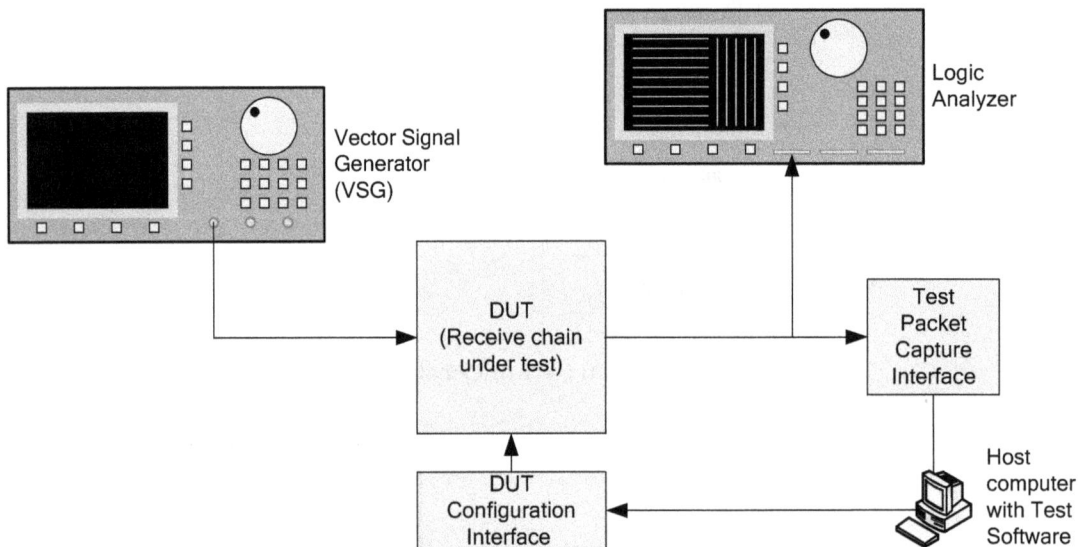

Figure 2.6: Example Receiver Test Setup

An FER measurement involves injecting a high-quality RF signal carrying Medium Access Control (MAC) frame data at a predefined signal level and measuring the ratio of errored frames to good frames at the output of the receiver. Signal injection is almost always done using a calibrated microwave signal generator that can modulate WLAN signals – a VSG – in order to obtain a high-quality signal. (In a pinch, another WLAN device can be used. This is referred to as the "golden radio" approach, but is not a recommended practice, as the signal quality of most off-the-shelf WLAN radios is both poor and highly variable.)

The WLAN system itself is normally used to capture the received frame data, as this is by far the simplest method; it is essential to ensure that the frame injection rate is kept low so that the WLAN system does not miss any frames. A special software program is executed on the WLAN system to measure and output the FER at the end of the test. The frame check sequence (FCS) in the received frames is commonly used to detect errors. A bit error in any portion of the MAC frame causes the FCS check to fail, and thus the FER is simply the ratio of the number of FCS failures detected to the total number of frames received. If the frame contents are known, however, it is also possible to detect errors by comparing the received frame data against the expected value. This has the added advantage of indicating when a frame has been corrupted by multiple errors.

2.5.5 System-level RF Measurements

In addition to performing device-level or subsystem-level RF tests, such as the ones already described, it is necessary to perform system level tests, as the whole WLAN assembly (receiver, transmitter, antenna, control CPU, software, etc.) interacts in such a way as to change the RF behavior of the system. For example, it is not uncommon to find radiated noise from the digital portions of a WLAN system (e.g., a laptop with a built-in WLAN client) coupling through the antenna into the RF portion, causing reduced receiver sensitivity and increased transmitter noise and spurious signals. Also, the WLAN antennas will interact with other metallic objects within the system, such as the cables and enclosures, causing a marked change in the radiation pattern and emission characteristics.

System-level RF tests are somewhat complex, as they involve a three-dimensional component to the tests (to account for the antenna radiation pattern) and must also be done in a high-quality anechoic chamber. An example of such a test, which measures the total radiated power (TRP) of a WLAN client, is shown in the figure below.

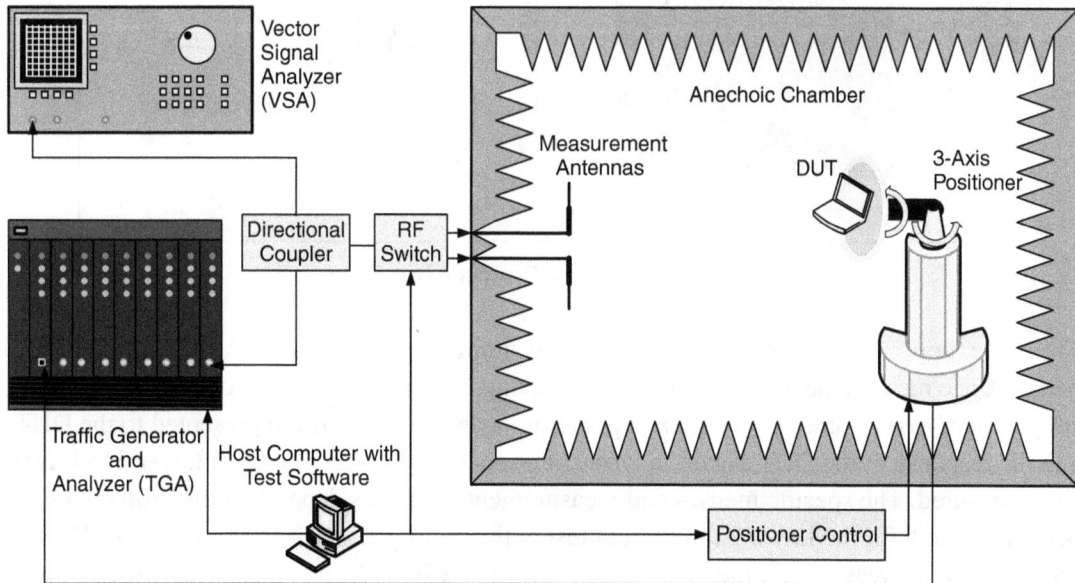

Figure 2.7: Example of a TRP Test Setup

In the above setup, a three-axis positioner is used to rotate the DUT about the X, Y, and Z axes while measurements are made. A fixed reference antenna picks up the signal transmitted by the DUT at each orientation, and this is then measured by the spectrum analyzer. The DUT is made to transmit frames at a fixed power and a reasonably high rate (in the case of clients, this is done with a software program running on the computer containing the client adapter). The result of the test is the total power radiated by the DUT as integrated over all directions – i.e., the Total Radiated Power or TRP.

2.5.6 Protocol-Level Testing

Once the RF characterization is complete, it becomes necessary to do the same for the digital and software subsystem. This type of testing is usually done using a clean and well-controlled RF environment (i.e., Layer 1 is essentially assumed to be working and factored out), focusing on the Layer 2 and higher protocol behavior. Note also that WLAN benchmark testing is normally performed in the form of protocol tests.

Almost all protocol-level performance and functionality tests are done using a frame generator and a protocol analyzer, typically combined into a single unit (sometimes referred to as a traffic generator and analyzer or TGA). An example of a typical protocol test setup, used for such metrics as throughput and latency, is shown in the following figure.

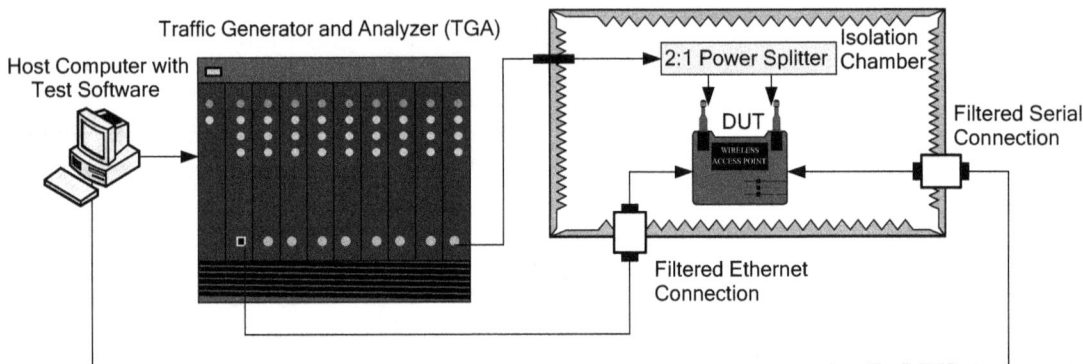

Figure 2.8: Example Protocol Test Setup

The general operation of the above setup is quite simple, even though the variety of tests and performance measurements can be quite bewildering. Controlled traffic streams ranging from simple MAC-level frames to complex application-layer transactions are presented to the DUT, and the response of the DUT (again in terms of protocol layers ranging from Layer 2 to Layer 7) is measured. The specific metrics and measurement processes depend on the protocol layer being activated; for the Layer 2 throughput test in this example, the traffic load on the DUT is progressively increased until the loss reaches some (small) predetermined threshold, at which point the offered traffic load is equal to the throughput.

WLANs exhibit considerable interaction between the RF layer and the higher protocol layers, which tends to blur the distinction between protocol tests and RF tests. WLAN TGAs therefore combine a few of the basic capabilities of signal generators, signal analyzers, and power meters in order to measure these interactions.

2.5.7 Environmental and Burn-in

Once the WLAN system or device has been shown to be working (essentially, proven to meet the characteristics promised on its data sheet), a final step of the process is environmental qualification: verifying that the device meets the radiated emissions, temperature, humidity, electrical isolation, safety, and long-term reliability requirements. For example, temperature and humidity tests are carried out in an environmental test chamber, as shown below.

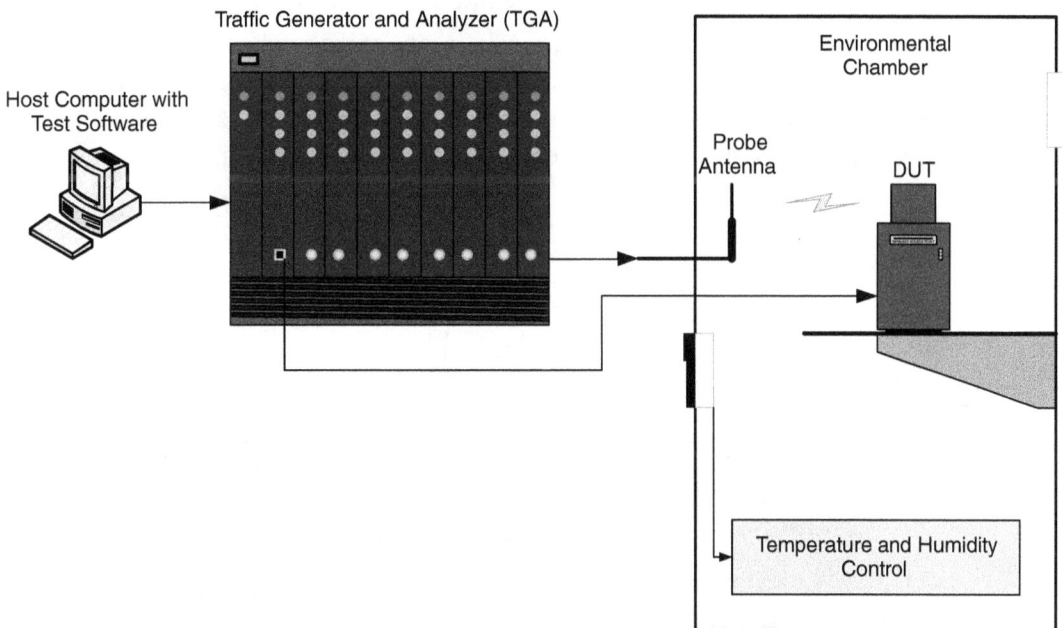

Figure 2.9: Typical Environmental Test Setup

Environmental tests using the above setup are relatively simple, though certainly time-consuming. The WLAN system/device is essentially placed in the chamber, powered on, and then exercised in some manner while the temperature and humidity are forced to a range of predetermined values as specified by the DUT datasheet. (For example, a typical WLAN device may have an operating temperature specification of $-20°C$ to $+55°C$, and a humidity specification of 10% to 90%.) The test is run for some period of time, usually ranging from 40 hours to several days; if the DUT continues to work normally during this entire period, it is considered to have passed.

2.5.8 Manufacturing

Manufacturing tests use much of the same test equipment as in the previously described test setup, but take a different tack: the emphasis is on verifying that a large number of copies of the device have been correctly constructed, rather than on testing basic functionality. It is not uncommon to find some development and QA tests being carried over to the manufacturing floor after suitable modification. A pass/fail indication is all that is required to filter out bad devices, and the tests themselves are simplified to keep test times short and manufacturing costs low. Every unnecessary minute spent on the manufacturing floor can translate to as much as $1.50 in extra cost for the final product, which makes quite an impact on the profitability of a product that can sell for under $50 in many cases.

Manufacturing test for WLANs is complicated by the fact that the radio must be tuned and calibrated before it can even begin to work. Test setups are commonly created on an ad hoc basis by the device manufacturer to satisfy the special needs of the products as well as to keep test times low. There is hence no standard or universally applicable test setup. Recently, however, some test equipment vendors (such as Agilent Technologies) are starting to offer a standardized arrangement for manufacturing test, an example of which is shown below.

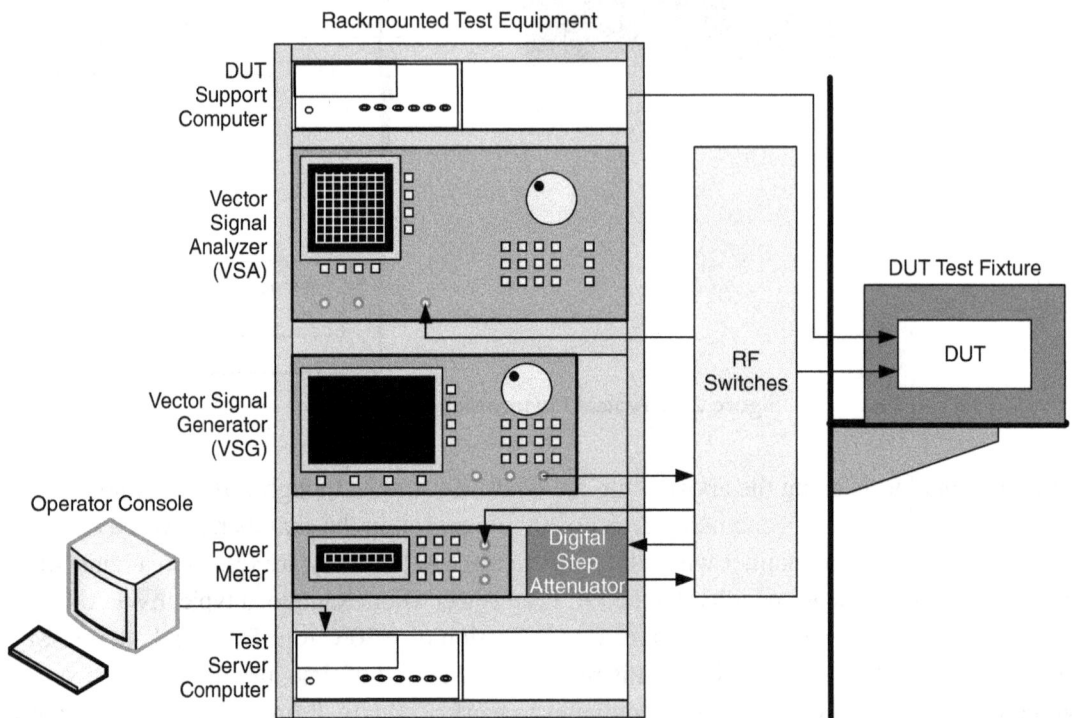

Figure 2.10: WLAN Manufacturing Test Setup Example

The manufacturing test setup above is built around a test fixture designed for quick insertion and removal of the DUT, while shielding it from interference. An RF power meter, a VSA, a VSG, a programmable power supply, a DC voltmeter and ammeter (for power consumption measurements), and a computer are connected to the test fixture. Various ancillary devices are not shown, such as the flash memory programmer that loads calibration constants into the flash memory present on the radio module. A separate control computer is used to drive the whole arrangement, running an automated program when prompted by the human operator after he or she has placed the DUT in the test fixture.

The test process is usually quite complex, even though (for obvious reasons) virtually none of this complexity is apparent to the operator. The operator merely enters a few simple parameters, such as the lot code and the MAC address of the device, and presses a button; the test system then takes over, calibrates the radio, runs a small battery of RF and protocol tests on the DUT, and finally presents a pass/fail indication to the operator. (Manufacturing engineers refer to this as the "happy face/sad face" output!) If the DUT passes, the operator sends it on to be packaged and shipped; otherwise, it is binned for failure analysis and rework.

2.5.9 Installation and Site Survey

Currently, WLAN installation setups and processes run the gamut from nothing at all ("install and pray") to systematic procedures involving propagation modeling, frequency planning, site surveys, spectrum analysis, trial networks, and pre- and post-install testing. Installation tends to be quite labor-intensive and often requires follow-up visits ("truck rolls") to clean up issues and fix unforeseen interference problems.

One of the most common installation practices is to conduct a site survey of reasonable complexity prior to actually placing APs and wiring them to the LAN infrastructure. The site survey is done by walking around the area in which the WLAN is to be installed. The installer carries some type of signal monitoring device or setup to measure the signal received from the AP as well as any interference from adjacent WLANs or other equipment such as microwave ovens.

Companies such as Berkeley Varitronics Systems (BVS) manufacture handheld equipment specifically for site surveys, that combines the functions of a signal monitor, a propagation analyzer, a basic spectrum analyzer, and a sniffer (protocol analyzer). In some cases a GPS unit may be integrated to automatically mark locations at which measurements are made, though in most commercial buildings made of concrete the use of GPS repeaters is usually necessary to get the GPS system to work. The site survey process with such equipment consists of placing one or more APs in strategic locations, generally as close to the anticipated final locations of the APs as possible, and then walking around the building or outdoor area marking the signal strength of both the test APs and interferers on a floorplan.

Figure 2.11: Site Survey Tools

Photo copyright © Berkeley Varitronix corp., provided by courtesy of Berkeley Varitronix corp.

After the floorplan has been comprehensively marked, a coverage map can then be created (using software tools available for the purpose) that indicates the signal strength that will be available in various areas. The proposed AP placements can be adjusted if necessary and the process repeated. Once a workable set of AP placements is obtained, the rest of the installation can proceed.

2.6 Repeatability

Unlike their wired counterparts, wireless test setups are characterized by a high susceptibility to noise and external interference, which can cause variability in both the input signal stimulus to the DUT and the response of the DUT to the input signal. This is particularly true if care is not taken with the test equipment as well as the test setup and the connections between various pieces of gear. The high variability usually shows up as a lack of repeatability between successive measurements; noise, interference, and signal level variations are random and tend to change over time. Another symptom of a poor test setup or problematic test equipment is a lack of reproducibility (i.e., repeatability between different test setups). Especially in the case of tests performed with off-the-shelf laptops as part of the test setup (instead of being part of the DUT), it is not uncommon to find test results that are impossible to reproduce, as the software loaded on the laptops significantly affect both the generated traffic as well as the measured results. This section describes some of the more common artifacts that plague wireless test setups, and how to deal with them.

2.6.1 Noise and Interference

After the problems of test signal generation and analysis have been dealt with, noise and externally generated interference are the next two largest stumbling blocks in repeatable WLAN testing.

Noise is, of course, inherent in all active devices in the signal paths (such as amplifiers and mixers) as well as resistive elements; it can thus be reduced but never wholly eliminated. Loose or corroded connections and cold solder joints are also sources of unwanted noise, particularly at the lower frequencies.

Interference is a larger problem. Not only do WLANs operate in unlicensed bands where there is a lot of RF activity ranging from cordless phones to video links, and unintentional radiators such as microwave ovens, but the proliferation of WLANs means that interference can be received from WLAN devices in adjacent networks, over which the user has no control. WLAN receivers are relatively sensitive and hence prone to such co-channel and adjacent-channel interference; it is not unusual to find a WLAN DUT being able to pick up and respond to signals from a different floor in the same building, or a different building altogether. The following table gives the minimum receiver sensitivity figures for various PHY bit rates, as specified by the IEEE 802.11 standard for 802.11g PHYs. Note that commonly available WLAN devices can perform at least 5–6 dB better than the minimum limits specified.

PHY Bit Rate	Minimum Sensitivity
Mb/s	dBm
6	−82
9	−81
12	−79
18	−77
24	−74
36	−70
48	− 66
54	−65

It is clear from the above table that the shielding effectiveness of enclosures or chambers must be quite high if external signals are not to interfere with the test. If the DUT responds to signal levels of −82 dBm, for example, and there is a nearby 2.4 GHz external source (such as an AP belonging to a different network) radiating at +20 dBm about 1 m (3 ft) from the DUT, then an enclosure with at least 84 dB of shielding effectiveness will be needed to prevent the DUT from being interfered with by the AP. (The RF field near the DUT from a 2.4 GHz +20 dBm omnidirectional source that is 1 m away will be about −8 dBm, and 10 dB or more of margin should be provided.)

In addition to a good chamber or enclosure, proper shielding techniques are necessary, particularly around the DUT or SUT. (Well-constructed test equipment uses metallic enclosures with electromagnetic interference (EMI) gaskets, and the incidence of both stray radiation and interference pickup from lab-quality test equipment is quite low.) It is therefore necessary to use RF shielded enclosures or chambers for all DUTs and ancillary unshielded equipment that forms part of the test setup. Note that even "non-RF" devices such as LAN switches that are connected to the DUT can become sources of interference problems if EMI filters are not interposed between the DUT and these devices; for example, signals picked up by the external device can be conducted through the Ethernet LAN cables and re-radiate within the enclosure or chamber containing the DUT. All elements of the test setup must be considered when improving shielding effectiveness.

An often overlooked factor in shielding effectiveness is the quality of the coaxial cables used to interconnect the RF portions of an otherwise well-shielded test setup. Good-quality SMA cables have shielding effectiveness of 110 dB or more if they are kept relatively short (1–2 m); they have multiple layers of copper braid surrounded by a continuous foil jacket, with no breaks in either braid or foil. Unfortunately they are also fairly expensive. There is thus the temptation to substitute lower-quality, less well-shielded cables, particularly for the longer runs where cable costs can become quite high. This can lead to severe interference issues, as poor-quality cables can in fact act like antennas, picking up everything within range of the DUT. Further, the high-loss dielectric and reduced braid coverage of poor-quality cables will attenuate the desired signal more (relative to the interfering signal), further exacerbating the problem.

Besides the above precautions with regard to shielding, it is best to avoid proximity to high-power devices when possible. Microwave ovens for instance, are notoriously leaky. Even when adhering to industry-standard shielding effectiveness requirements, a 600 W microwave oven in operation emits over 0.5 W of 2.4 GHz energy, which appears as a broad, noisy, 60 Hz modulated carrier. Even an 80 dB isolation chamber is unable to prevent the output of a microwave oven from drowning out WLAN signals that would otherwise be received quite well. (The 80 dB of chamber isolation reduces a 0.5 W stray emission to only −53 dBm, which is well above the typical WLAN receiver sensitivity threshold.)

Care must be taken even with open-air setups, where (presumably) shielding around the DUT is not required. The stray emissions from internal elements of the DUT – for example, the CPU and auxiliary digital chipset devices surrounding it – and its harmonics can easily fall into the 2.4 or 5 GHz bands. This can be radiated through the DUT enclosure to affect other components of the test setup (e.g., other elements of the SUT). As stray emissions rarely produce a coherent signal such as a strong, narrow carrier, they are not readily distinguished on a spectrum analyzer and hence can be an insidious problem in unshielded test setups. A solution, of course, is to place the DUT in a chamber and cable its antenna connector(s) to an external reference antenna.

2.6.2 *Controlling the Stimulus*

It is unfortunately common to find test and QA engineers being aware of the issues of interference and going to great lengths to exclude it, but then completely neglecting another major source of repeatability issues: control of the test stimulus.

In standard RF testing, the source of the test stimulus (i.e., the signal source) can obviously cause many issues with the test results if it is not properly selected and calibrated. This is fairly well known; few engineers would choose to use a homemade or "hobby" signal source when they need to obtain consistently accurate and repeatable results. A good-quality VSG with NIST-traceable calibration is usually essential.

However, when the level of testing rises above the PHY layer, it is not unusual to discover that the VSG has been replaced by a desktop or laptop computer running some sort of software program. For simplicity (and to save on time and money), an off-the-shelf WLAN client card of some sort is incorporated, because these come with standard drivers and protocol stacks for most widely used operating systems and further reduce the work the engineer has to do in terms of getting the test setup running. The result is then used to generate test signals, traffic streams for performance or interoperability testing, and all manner of other signal stimuli. (Sometimes the "software program" used to produce the traffic may be nothing more than a "drag-and-drop" of a file via the OS GUI!)

Unfortunately neither the laptop, the WLAN client hardware, nor the driver and OS have been designed as test equipment. Consumer-quality devices in fact cut as many corners as possible in order to reduce both time-to-market and cost. The manufacturing tolerances on the transmitter signal level of the WLAN client's radio may have as much as a 6 dB range, if the signal levels are specified at all. The problems with the driver and OS are even worse; it is not uncommon to find a client laptop completely changing its wireless "personality" when the driver is updated, as the driver plays a significant role in the behavior of the WLAN adapter, from controlling the output power to inserting gaps in the transmitted frame stream.

While it is possible to gain some control of – or at least some knowledge of – the power output and data transfer rate of a WLAN laptop client being used as a signal source, it is far better to use a VSG or a dedicated hardware traffic generator when repeatable tests are to be performed.

2.6.3 Common Errors

Most common WLAN test setup issues, in addition to the above, can be solved by the application of simple common sense. Some errors are, however, less obvious than others; for example:

a. Poor terminations.

    It is not uncommon to find a 3:1 or 4:1 splitter being used when a 2:1 splitter is required, simply because it was conveniently at hand. There is nothing particularly egregious about this practice, as long as the extra loss is not a factor; however, the unused ports on the splitter must be terminated with good-quality 50-ohm terminators. Leaving SMA ports open on any device (not just splitters) in the signal path leads to hard-to-find problems with reflections as well as stray pickup on the unterminated ports. In particular, WLAN PAs subjected to reflections from unterminated ports can cause all manners of unrepeatable effects, including instability and oscillations.

b. Poor connector discipline.

    Continuity of shielding and avoidance of impedance bumps is important, especially at the connectors; typical SMA connectors have greater than 100 dB of shielding effectiveness and less than 0.1 dB of return loss, but only if they are properly installed and assembled. (The torque wrench has a purpose!) A badly torqued SMA connector exhibits either a drastic increase in return loss due to compression of the dielectric, or a reduction in shielding effectiveness due to a discontinuous shield. A loose SMA connector can add 10–20 dB of intermittent attenuation in the signal path due to a poorly mating center conductor, so that when the cable is wiggled the test results vary by large amounts. Also, the gender of SMA connectors should be carefully matched; a reverse-SMA antenna or cable connector will fit quite happily on a standard SMA bulkhead jack, leading to a roughly 50 dB inline signal loss because the center conductor of the cable cannot make contact with the center conductor of the jack.

c. Regarding DUT enclosures as shielded, without testing them first.

    APs are sometimes packaged in all-metal enclosures, and there is a temptation to assume that no chamber is required. The shielding effectiveness should be verified by experiment before trusting to it, because while the enclosure may be metallic and well made, the power, LAN, and serial console connectors are likely to have inadequate filtering. This is not too surprising considering that the attenuation of standard EMI filters drops drastically above 1 GHz (in fact, it is unusual to find any that are specified above 2 GHz). Poorly filtered AC or DC power cords and unshielded LAN cables act as antennas, conducting WLAN signals into or out of the DUT enclosure with very little attenuation.

d. Poor grounding.

    Wired LAN test setups can generally ignore grounding issues completely: Ethernet cables are transformer coupled with guaranteed isolation levels of up to 1500 V, and optical fibers of course are non-conductive. Thus test equipment can be connected up to the DUT in any random pattern without concern for ground loops or ground potential differences. Over-the-air WLAN test setups are just as good in terms of freedom from ground loops, but cabled WLAN setups are another matter entirely. First, the RF shields of the cables forms low-impedance paths between pieces of equipment, and thus ground loops are a very real possibility. Second, the center conductors of these cables form paths for AC and DC currents between signal ports, causing ground potential differences to play havoc with test results. It is advisable to power all pieces of equipment from a single AC outlet if possible; failing that, suspect ground loops when bizarre or inexplicable failure behavior is observed.

WLAN Test Environments

Wireless LAN (WLAN) equipment is tested in various types of environments: open-air, chambered (screened room or anechoic), or conducted. The environment selected has a significant impact on what types of tests can be done and what type of test instruments are used. This chapter treats the different test environment types and their characteristics. Also, it describes when each type of environment is used, and what problems are faced by people using each type of environment. Repeatability of measurements will be a key focus of this chapter.

3.1 Wired vs. Wireless

Wired LAN testing rarely bothers with specifying any form of physical environment. Cables and optical fibers work equally well whether coiled up in a tangle or stretched out behind walls and ceilings. Hubs, switches, routers, and computers do not care whether they are mounted vertically or horizontally. The materials used to construct the walls and ceilings do not have any impact on the test results. Wired LANs can be run without noticeable impact in close proximity to most electronic devices – optical fiber LANs can be run in even extremely electronics-hostile environments without degradation.

WLAN performance, on the other hand, is significantly affected by changes in the environment. For example, place a large metal can over an access point (AP); the connectivity between the AP and a nearby client will fail immediately. The distance between the AP and its clients is also significant. Things work much better in open-air or low-metal environments (e.g., homes made of wood and wallboard) rather than buildings with lots of metal in the walls and ceilings. Turning on a microwave oven next to an AP in the 2.4 GHz band usually results in a sharp increase in interference and a drop in performance.

It is hence necessary to consider the physical environment surrounding the equipment being tested as part of the test setup, and to incorporate its characteristics into the test methodology, in order to successfully test WLAN equipment.

3.1.1 Terminology

The following terminology is defined to aid in understanding the descriptions of the various test environments that follow.

Conducted (or cabled): The RF signal is carried by means of cables (or waveguides, though this is quite rare). Antennas are typically not used.

Near-field probe: A small "antenna" that is used to couple test signals to and from the antenna(s) of the Device Under Test (DUT), particularly in situations where the DUT antenna(s) are not removable. Cables are then used to interface the test equipment to the near-field probe.

Over-the-air (OTA): The RF signal is propagated at least some distance through the air, with no metallic connection between transmitter and receiver.

Indoor: The environment existing inside a building such as a home or an office, such that the RF signal encounters walls and ceilings in its path.

Outdoor: A environment with an open space in which both transmitter and receiver are placed; the RF signal usually has a direct line-of-sight path with no intervening walls or ceilings, though reflections may occur from distant objects or buildings.

Controlled: A situation where the physical properties of the environment (e.g., its size, metallic objects, absorbers, etc.) can be determined, and kept unchanged.

Uncontrolled: The opposite of controlled, in that the properties of the environment either cannot be determined, or cannot be kept constant. For example, an office in daily use with people and objects moving around it during the normal workday is an uncontrolled environment.

Open-air: A test environment where both the transmitter and receiver are placed in the open (i.e., not cabled or isolated) and RF signals propagate between the transmitter and receiver OTA. Open-air environments are prone to interference.*

Chambered: A test environment where a metallic enclosure is used to confine the DUT (and possibly the test equipment as well) in order to exclude external interference and noise.

Isolation: The degree of attenuation of unwanted signals (e.g., external interference and noise) by a metallic enclosure or shielded cables.

3.2 Types of Environments

3.2.1 Open Air

Open-air environments use the actual physical surroundings (or a replica thereof) in which the WLAN equipment will be deployed. In some cases, an open-air environment may be an

---

*Note that there is constant confusion between the use of "over-the-air" and "open-air" as used in common industry practice. In this book, an "open-air" setup is considered to be a subset of an "over-the-air" set up. "Open-air" for our purposes indicates an unshielded, "real-world" scenario; "over-the-air" refers to any situation, including a fully chambered set up, where the RF signal propagates without a cable between the transmitter and receiver.

antenna test range, which is specially designed to be free of reflectors and absorbers. It is also not uncommon to find several rooms or even an entire floor of a building being used as a WLAN test environment. Some of the larger WLAN chipset or equipment vendors dedicate space for the purpose, rather than making use of active office buildings, in order to gain some measure of controllability over the environment.

Open-air environments are principally used for the following kinds of tests:

1. Antenna testing, for determining antenna radiation patterns, total radiated power, total isotropic sensitivity (see Section 4.3.4).

2. Evaluating multipath effects on receivers and transmitters, particularly in the case of Multiple Input Multiple Output (MIMO) systems.

3. Large-scale equipment performance evaluation (place the equipment in a building and then determine how well it performs according to different metrics).

4. Pre-deployment performance testing of WLAN equipment in situ.

Dedicated open-air environments are normally very expensive (as the cost of renting an office or a building floor is part of a test setup). However, for certain scenarios this may be the only method of carrying out the tests. An example is in evaluating the performance of a full-scale WLAN, especially in terms of understanding the impact of the building walls and ceilings on the system, and also including the external interference sources that may be encountered.

3.2.2 Screened Room

A screened room is a large room made of metal (usually copper) mesh or sheet into which the test equipment, the System Under Test (SUT), and the operator are placed. Screened rooms act as Faraday cages; the name "Faraday Cage" derives from the effect discovered by Michael Faraday wherein RF currents flow on the outside surface of an object, penetrating very little into the object itself. A screened room thus provides excellent isolation of interference and noise. It also confines RF signals generated by the DUT or test equipment to the interior of the cage, and keeps them from interfering with external equipment – a significant factor at high RF powers.

Screened rooms are used for system-level performance tests where it is inconvenient or impractical to cable up the DUT(s) and tester, but it is essential to exclude interference. For example, testing a laptop with a built-in wireless interface may require access to the display and keyboard in order to configure and control it. In this case, the laptop, an AP, and the test equipment are all brought into a screened room, and the tests carried out there.

Screened rooms are quite expensive – a room-sized cage has to be built out of copper sheet or mesh, and provided with adequate power and ventilation. All joints in the cage must be properly constructed to eliminate leakage. Further, a door is necessary for the operator to go in and out, and to bring in equipment. Ensuring that the door seals do not leak, even after long periods of use, is a difficult task.

Screened rooms are not generally suitable for OTA testing of WLAN devices as their walls are made of bare metal mesh or sheet and hence they produce tremendous amounts of multipath. (They act more like reverberation chambers in this regard; reverberation chambers will be discussed later.) An isolated environment that is suitable for OTA testing is the large anechoic chamber, which has metallic inner walls that are completely covered with specially shaped RF absorbing material.

3.2.3 Conducted

A conducted environment is in some sense diametrically opposite to both an open-air set up and a screened room. In this situation, the antennas of the DUT (as well as the tester, if it has any) are removed, and well-shielded cables are used to interconnect the DUT and the rest of the test equipment. All RF paths are thus not only isolated but also confined to shielded cables. This not only excludes interference, but allows tight control of the attenuation and amplitude/phase properties of the signal paths. It is also assumed that the DUT and test equipment are inherently shielded by their enclosures; if this is not true, then a chambered environment must be used instead.

Conducted setups are used for the majority of all WLAN testing, from low-level RF measurements to application layer performance measurements. They are reasonably low-cost, easy to set up and use, compact and flexible. Distances in the real world can be emulated using attenuators or power level control. Channel characteristics (multipath, etc.) can even be emulated, albeit somewhat expensively, using channel simulators.

Conducted environments cannot be used for two kinds of WLAN test scenarios: where the radiation patterns of the antennas mounted on the DUTs is a factor (such as measurements of the antenna patterns themselves), and where the open-air interaction between many APs and clients is a factor. The first limitation is fairly obvious – removal or bypassing of the antennas removes any possibility of measuring their impact. The second is less clear. To some extent, a small number of WLAN devices can be interconnected with cables and carefully selected attenuators to simulate a real deployment. However, if a lot of devices must be interconnected, the setup becomes quite complex and unmanageable. Further, if the variations in attenuation between different devices become large, RF sneak paths are set up that effectively place a lower limit on the attenuation factors, preventing close simulation of the target network.

3.2.4 Chambered

A chambered test environment is midway between a screened room and a fully conducted setup. It is also a Faraday cage, and comprises one or more isolation chambers into which the

DUT(s) are placed; cables are then used to interconnect the DUT(s) and the test equipment. In the case of DUTs with built-in antennas, near-field probes can be used to couple RF signals into the antennas. All other aspects of chambered testing follows that of a fully conducted environment, including the use of attenuators or transmit power control to simulate distance. The isolation chambers act as miniature screened rooms, excluding external interference and preventing the DUTs from interacting with each other.

A chambered test setup has several advantages for most WLAN testing. It is much less expensive and space intensive than screened rooms; typical chamber dimensions are measured in inches, as compared to tens of feet for screened rooms. In fact, a chamber need not be much bigger than the DUT it contains. It eliminates the problem of poor enclosure isolation in the DUTs, particularly those with metalized plastic casings, that is a problem with fully conducted environments. Finally, it is the only feasible way of isolating DUTs with built-in antennas, when the test setup must contain a large number of such DUTs.

The principal disadvantage of isolation chambers, apart from higher cost and bulk (as compared to conducted tests) is the need to use door seals and filters. Most isolation chambers are equipped with doors that may be opened to allow access to the DUT. The door must be adequately sealed when closed to prevent RF leakage from ruining the isolation. This requires a flexible gasketing material of some sort that forms an RF-tight seal when the door is latched shut. Flexible door gaskets, however, tend to degrade over time. For instance, the elastomer-cored metallic braid used in inexpensive RF gaskets takes on a "set" when compressed for a while, and eventually leaves a small gap; even a gap of a few hundredths of an inch can reduce isolation by 20–50 dB, enough to ruin an accurate performance measurement.

Filters are another problem with isolation chambers. The DUT must be supplied with power, and also (in the case of APs) needs some sort of wired Ethernet interface so that test data can be sent and received. Thus cables must be brought in and out of the chamber. To prevent these cables from acting as antennas, low-pass filters are inserted at the points where the cables cross the chamber walls (bulkheads). The design of filters that can keep RF out while still enabling high-speed Ethernet data to pass is not trivial. In some cases, the problem is sidestepped by using an optical Ethernet link rather than twisted pair.

3.3 Outdoor and Indoor OTA

OTA environments are divided into two types: outdoor and indoor. Each type has its own issues that must be considered.

3.3.1 The Outdoor Environment

The outdoor environment is typically a large open space; the location and size of the space is usually dictated more by the access to sufficient land, rather than being the best choice of the designer. For example, when rolling out a WLAN hot-spot in an airport, it is unlikely that the designer will have the luxury of testing the equipment in a fully functioning airport! Instead, some reasonable approximation to the target environment is set up, the tests are performed, and the results are extrapolated to predict how the equipment will function in the ultimate deployment.

An antenna range is a special case of an outdoor environment, consisting of a relatively large and flat area from which all unwanted obstructions and interference sources have been removed. Only the conductivity of the ground remains as a factor; this is calibrated out using measurements made on reference antennas at known power settings. (An exception is a reflection range, where the ground forms part of the measurement setup and specular reflection from the ground combines constructively with the direct rays to create a uniform field – the quiet zone – about the DUT.) As implied by the name, antenna ranges are most often used for testing and characterizing the efficiency and radiation patterns of antennas, especially in the VHF, UHF, and low-microwave frequencies. However, with the increasing interest in sophisticated beamforming and multiuser detection schemes in wireless data networks, antenna ranges may be used to test complete systems as well.

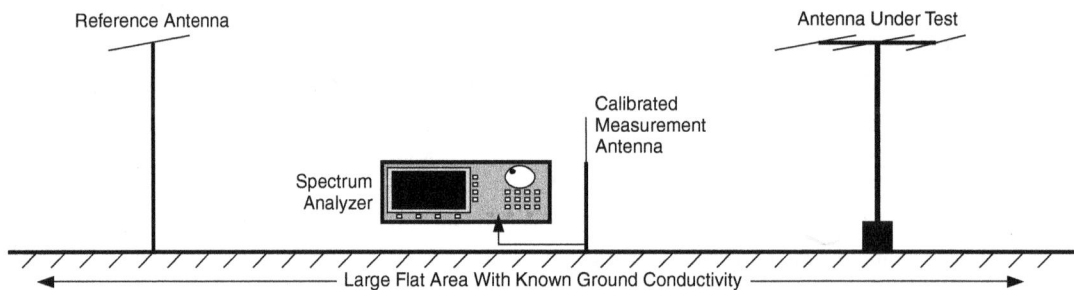

Figure 3.1: An Antenna Range

3.3.2 Indoor Environments

Indoor environments are basically various types of buildings, such as homes, offices, warehouses, factories, etc. Every indoor environment is basically unique, especially at microwave frequencies. The composition of walls, ceilings and floors, the presence of metallic objects, diffracting elements such as metal wall studs, etc. all play a role in propagation. Indoor environments should thus be considered 'one-off' and irreproducible.

For example, at frequencies of 2.4 GHz and above, every metallic object on the order of 10 cm (4 in.) or larger becomes a scatterer of electromagnetic waves; hence the number and position of even innocuous objects such as computers and filing cabinets significantly affect the environment "seen" by WLAN receivers and transmitters. There is thus no such thing as a "standard" or "common" indoor environment. Instead, vendors test in as many buildings as they find feasible, and extrapolate the results. Many WLAN vendors simply test their devices within their own cubicle farms or borrow a house to test consumer products.

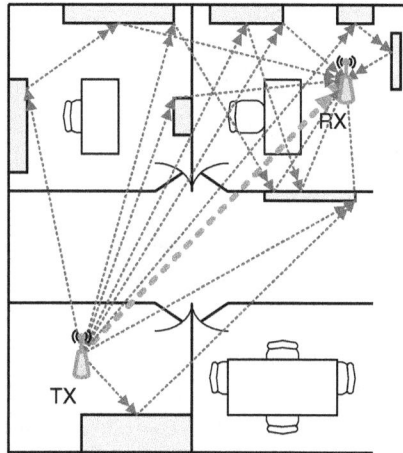

Figure 3.2: The Indoor WLAN Channel

3.3.3 Fresnel Zones

A key parameter in most over-the-air (OTA) test scenarios, is the diameter of the first Fresnel zone between the transmitter and receiver. Fresnel zones are a series of imaginary concentric ellipsoids (ellipses of revolution), with the transmitter and receiver at the foci of each ellipse. The ellipses are therefore aligned with the line-of-sight path between the transmitter and receiver. The ellipse dimensions are such that each successive ellipse represents a region in which multipath rays must travel up to a half wavelength ($\lambda/2$) more than in the preceding, inner, ellipse. The following figure represents the first Fresnel zone.[1]

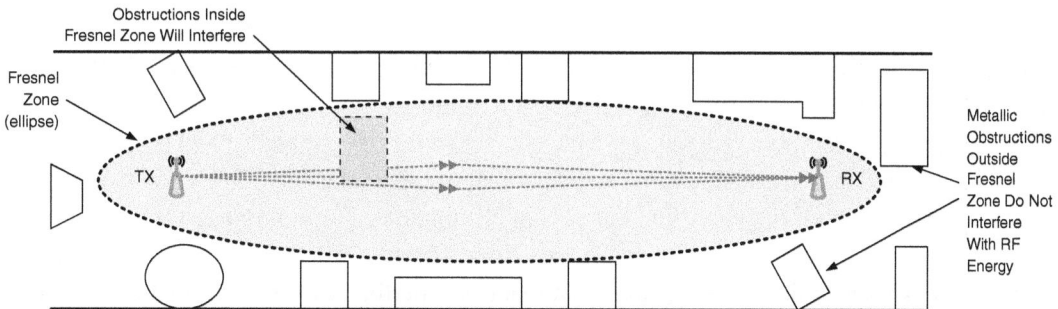

Figure 3.3: First Fresnel Zone

---

[1] Fresnel zones originally applied to point-to-point microwave systems, but also apply to indoor areas.

Objects within the first Fresnel zone will affect line-of-sight propagation between the transmitter and receiver; objects lying outside will not affect line-of-sight propagation (but can act as reflectors and hence induce multipath). Determining whether the Fresnel zone is unobstructed or not can thus also indicate whether the propagation model is Rayleigh or Rice. (The nature of Rayleigh and Rice propagation statistics is described in Chapter 9.) The general rule of thumb is that if at least 60% of the first Fresnel zone is clear of metallic obstructions, then the propagation statistics will adhere to the line-of-sight case.

3.3.4 Interference and Fading

Open air test environments are highly subject to variation due to interference and fading. Interference is caused by other devices operating in the vicinity, such as 2.4 GHz cordless phones, Bluetooth devices, other WLAN equipment, microwave ovens, medical diathermy devices, electromagnetic interference from computers and cell phones, etc. Windows and interior walls in office buildings often attenuate RF energy very little, particularly in the 2.4 GHz WLAN band, and hence the interference may originate from adjacent offices or even adjacent buildings. It is very difficult to completely eliminate interference from indoor environments.

Virtually all indoor propagation environments cause RF signals to bounce (reflect) off or bend (diffract) around obstacles, and thus take multiple paths to their destination. These multiple paths, which are usually of different lengths, are referred to as multipath. Fading is generally caused by the reflected or diffracted multipath signals combining constructively or destructively. Constructive multipath occurs when the complex amplitudes of the different multipath signals are in-phase (due to the path length differentials being just right) and add; destructive multipath when complex amplitudes are out of phase and cancel. Other fading effects are shadowing, which is a reduction in signal strength behind a large metallic obstruction such as an elevator shaft, and flat fading or slow fading, which is a gradual fall off of signal strength with distance.

Both interference and fading are subject to spatial and temporal variations. Movements of either the WLAN receiver or the transmitter by a few inches, (i.e., a small amount relative to the wavelength), or rotation by a few degrees, can cause significant changes in signal strengths of both the desired signal and any interference. Diversity antennas on WLAN devices can help to mitigate some of the variation, but cannot eliminate it. Further, unless the environment is completely static, the signal strength can vary by as much as 10–30 dB at intervals of a few seconds. This is typically due to the movement of metallic objects and human beings in and around the test environment, which in turn causes variations in the amplitude and phase of the reflected signals. In both cases, a constructive combination of signals can change to a destructive combination, and vice versa.

Dealing with the spatial and temporal variations found in both indoor and outdoor environments is key to any sort of meaningful OTA measurement. The temporal component may be handled by performing a number of measurements over a period of time and taking the average. In some cases a smoothed running average is taken over time to obtain a profile of both interference and fading. The spatial issues are usually dealt with by using turntables and/or linear positioning devices to physically move the receiver or transmitter (or both), with measurements being taken at each angle or location and then averaged. The movement may be regular (e.g., rotating the DUT in uniform steps of 10°) or random; mathematical analysis shows that the random steps may actually work better.

3.3.5 Characterizing the Indoor Environment

As previously mentioned, indoor environments are extremely variable in nature, with all sorts of unpredictable scattering and interference effects. It is therefore advisable to characterize them in some way prior to beginning a test.

A spectrum analyzer may be used to determine the amount of unwanted interference present near the WLAN receivers. Spectrum analyzers can indicate the rough nature of the interference pattern as well as its strength. Typical hand-held spectrum analyzers used for this purpose cost about US$10,000 and can be fitted with antennas for picking up interfering signals. It is usual to continuously scan for unwanted signals during the whole test process, due to the temporal variations in interference signals. If interference is found, the test may have to be repeated when it goes away.

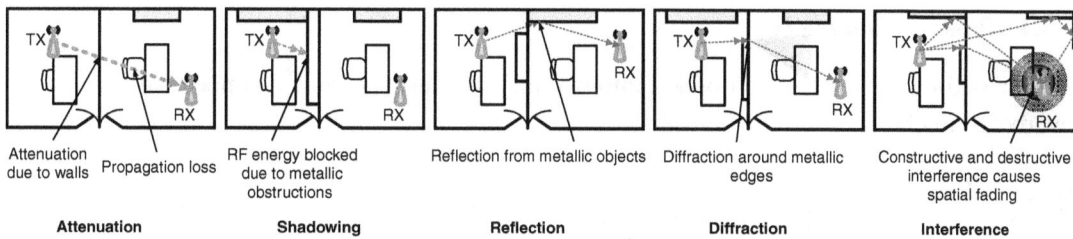

Attenuation due to walls	Propagation loss	RF energy blocked due to metallic obstructions	Reflection from metallic objects	Diffraction around metallic edges	Constructive and destructive interference causes spatial fading
Attenuation		**Shadowing**	**Reflection**	**Diffraction**	**Interference**

Figure 3.4: Indoor Environment Artifacts

Characterizing the multipath in the environment is more difficult. Propagation prediction software (e.g., the Motorola LANPlanner modeling package) can model the multipath and predict the resulting propagation characteristics quite accurately, provided that they are fed with a correspondingly accurate description of the environment. This includes floorplans, wall and floor compositions, locations and sizes of large metallic objects, etc.

In many situations, however, such detailed information is not available to be fed into a propagation modeler. In these cases, the only recourse is to make in-situ measurements using some form of channel sounder. A channel sounder acts much like a network analyzer, but directly measures the propagation characteristics of an RF channel rather than an RF device. However, their use is complex and time-consuming; thus most WLAN measurements fall back to the averaging scheme previously described.

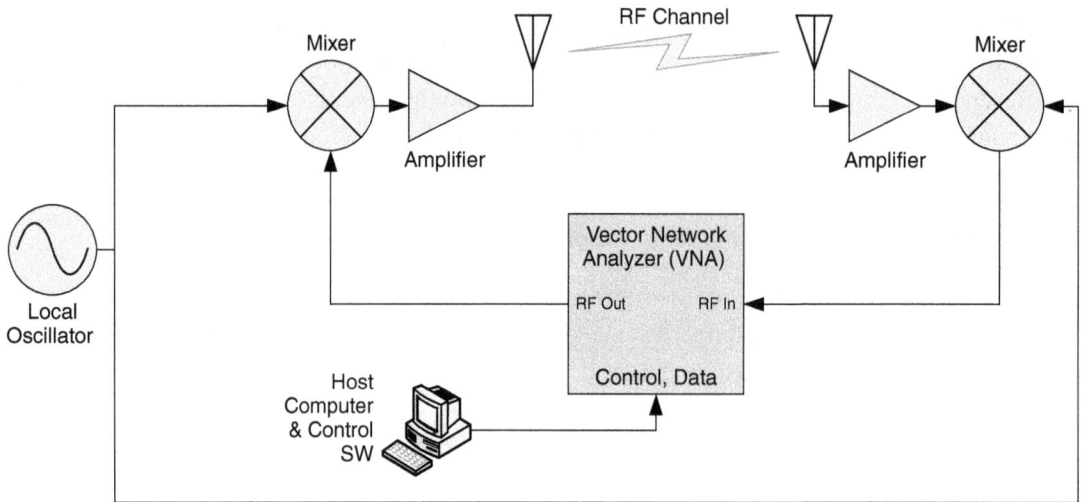

Figure 3.5: Typical Channel Sounder

Channel sounders will be described in more detail in Chapter 9.

3.4 Chambered OTA Testing

3.4.1 Types of Chambers

Chambers for OTA testing are divided into two major types: anechoic chambers and reverberation chambers.

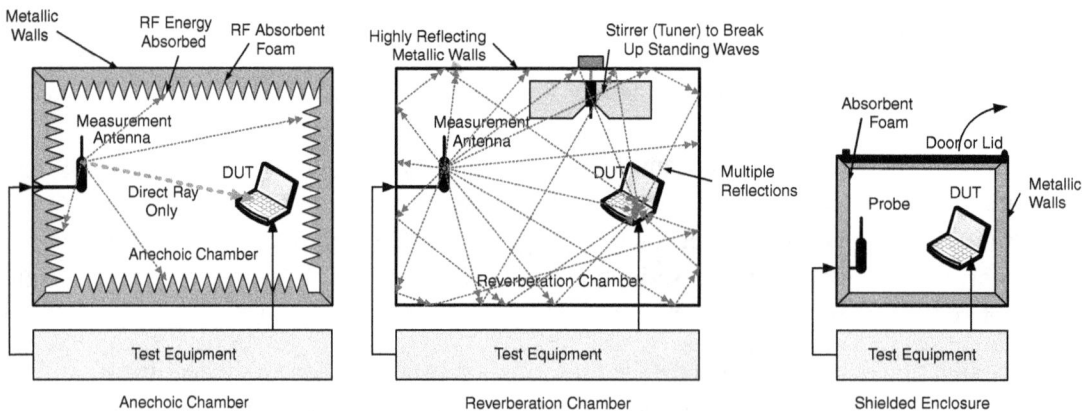

Figure 3.6: Chambered Test Setups

Anechoic chambers have walls that are lined with an absorbent foam material that minimizes the reflection of RF energy within the chamber. In larger anechoic chambers this material may be formed into wedge and pyramid shapes, so that any residual reflection from the surface of the foam is directed away from the DUT and eventually absorbed by some other portion of the foam. Anechoic chambers, or variants thereof, are used in most chambered tests, such as radiated power and sensitivity measurements or antenna patterns. Anechoic chambers are commonly rectangular and intended to simulate free-space conditions by maximizing the size of the quiet zone (described below). A variation is referred to as a taper chamber, and uses specular reflections from a pyramidal horn to produce a plane resultant wavefront at the DUT; taper chambers are used more commonly at lower frequencies.

Reverberation chambers are the opposite of anechoic chambers; they have no absorbent foam and are designed to *maximize* reflections. A "stirrer" or "tuner" is used to further break up standing waves that may form at specific frequencies within the chamber. The DUT is therefore subjected to a relatively uniform (isotropic) electromagnetic field with a statistically uniform and randomly polarized field within a large portion of the chamber volume. As the field is entirely confined within the chamber, the field density is also much larger than in an anechoic chamber or open-air site. Reverberation chambers are hence very useful for measurements of shielding effectiveness of DUT enclosures, rapid measurements of emissions or sensitivity covering all angles and polarizations, and so on.

A variant of an anechoic chamber is the small shielded enclosure whose main purpose is to exclude external electromagnetic interference from reaching the DUT. In this case the absorption of the walls within the chamber is of no consequence; however, for improved shielding and reduced self-interference (due to incidental radiation from the DUT internals) a thin layer of foam may be applied to the walls.

3.4.2 Far Field, Near Field, and Reactive Near Field

The space within an anechoic chamber is categorized into three zones: the far field, the radiating near field, and the reactive near field. The boundaries of these zones are located at different radial distances from the DUT, as shown in the following figure.

The far field (sometimes referred to as the "radiating far field") is the region in which electromagnetic energy propagates as plane waves (i.e., the wavefronts are parallel planes of constant amplitude), and direct coupling to the DUT is negligible. The shape of the field pattern is independent of the distance from the radiator. The radiating near field is between the reactive near field and the far field; in this zone, both electromagnetic (EM) wave propagation and direct coupling to the DUT are significant, and the shape of the field pattern generally depends on the distance.

The region closest to the DUT forms the "reactive near field"; this region is occupied by the stored energy of the antenna's electric and magnetic fields, and the wave propagation

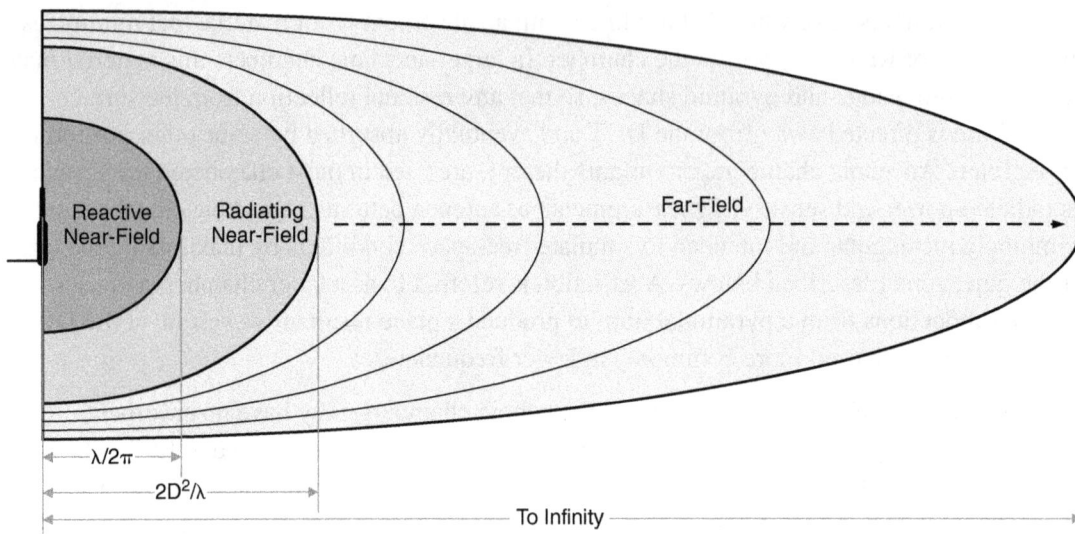

Figure 3.7: Near and Far Field

aspects are not significant. Direct inductive and capacitive coupling predominates here. Care must be taken with the size and placement of conductive (metallic) objects in the reactive near field, as they will couple and re-radiate considerable amounts of energy from the DUT; essentially, they become parasitic antenna elements. This materially alters the radiation pattern of the DUT antenna and causes substantial changes in the RF performance of the DUT. For example, power or signal cables running through the reactive near field of the DUT antenna can become parasitic elements and not only change the radiation pattern but also conduct and propagate RF energy to unexpected locations.

The distance from the DUT to the boundary separating the radiating near field from the far field is known as the Fraunhofer distance (or Fraunhofer radius), and is a function of both the wavelength used as well as the physical dimensions of the conductive elements of the DUT and the test equipment or test antenna. The Fraunhofer distance is given by the following equation:

$$R = 2\,D^2/\lambda$$

where R is the Fraunhofer distance, D is the largest dimension of the transmit antenna, and λ is the wavelength. For example, the Fraunhofer distance of a $5/8\lambda$ vertical radiator (antenna) at 2.4 GHz is approximately 10 cm (about 4 in.).

The Fraunhofer distance is not controlled purely by the RF antenna(s) of the DUT or test equipment. As previously noted, metallic objects within the reactive near field of the DUT will pick up and re-radiate RF energy; for example, in the case of a laptop, the WLAN antennas built into the laptop will induce circulating currents in the frame, metallic sheets, heat sinks, and hard disk enclosures, all of which will then act as antennas in their own right. The *D* component in the Fraunhofer distance is therefore nearly the width of the entire laptop, as much as 35 cm (14 in.). Obviously this can make the boundary between near and far fields extend much further out (in this example, 2 m, or about 6 feet), and the anechoic chamber may have to be made quite large to deal with this effect.

In most anechoic chambers of reasonable size, the DUT or test antenna can be moved within a small region without appreciably altering the energy level induced in the test antenna. This region is known as the quiet zone, and is caused by constructive interference of reflections from the walls of the chamber, thereby "smoothing out" the signal intensity over a small region. The size and shape of the quiet zone of an anechoic chamber is significantly affected by the construction of the chamber, and is always less than the Fraunhofer distance; hence determination of the quiet zone is done by experiment. The quiet zone is useful in that it reduces the need for extremely precise positioning of the DUT or test antennas for different test runs; all that is necessary is to locate them somewhere in their respective quiet zones. Vendors of anechoic chambers usually provide specifications of the size and location of the quiet zone in their products, or else it can be experimentally determined by moving a probe antenna around the chamber (with the measurement antenna being driven with a signal generator) and using a spectrum analyzer to indicate a region of minimum standing waves.

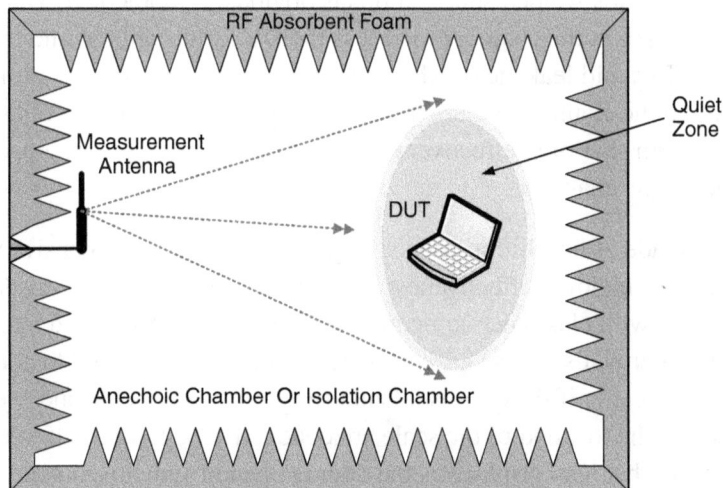

Figure 3.8: Quiet Zone

Note that the foregoing discussion of near field and far field is mostly relevant only to large anechoic chambers used for antenna pattern, radiated power, and receiver sensitivity measurements. In the case of small shielded enclosures, the chamber walls are well within the near field.

3.4.3 Coupling to the DUT

In all but the largest anechoic chambers the DUT is placed completely within the chamber but the test equipment and operator are located outside. In this case, the signals from the DUT are picked up via calibrated reference antennas. These are usually simple dipoles or standard ground-plane vertical radiators, that have been built to have as uniform a pattern as possible, and then calibrated using a signal source and a field strength meter or spectrum analyzer. Calibration is done in three dimensions so that the actual radiation pattern of the reference antenna is known and can be factored out of the measurements on the DUT.

In the case of small shielded enclosures, coupling to the DUT is done either directly (i.e., cabled to the DUT antenna jacks), or via near-field pickup probes placed very near the DUT antennas. Calibration of pickup probes is usually quite difficult and generally not performed. Instead, the DUT's own Received Signal Strength Indication (RSSI) report may be used as a rough indicator of the amount of power being coupled from the test equipment to the DUT.

3.4.4 Shielding effectiveness

The most complex and failure-prone portion of an anechoic chamber are the door seals. Perfect shielding is possible if the DUT can be placed permanently into a superconducting metal box with all the walls welded shut, but this is obviously impractical. Thus some kind of door must be provided in the chamber to access the DUT, the cabling, and the reference antennas or probes. To avoid leakage of RF energy from around the door (via small gaps between the door and the chamber walls), it is necessary to provide an RF-tight gasket to seal the gaps. The long-term shielding effectiveness of this gasket often sets an upper limit on the isolation provided by the chamber.

The gasket may be made of beryllium–copper finger stock, which is expensive but is very durable and has a high shielding effectiveness. Less costly gaskets may be made of woven wire mesh tubes filled with elastomer compound for elasticity, and inserted into channels in the door and chamber walls. These gaskets work well initially but tend to compress and deform over time, especially if the door is latched shut for long periods, and eventually lose their effectiveness. At the low end of the scale are conductive or coated self-stick gaskets, which not only deform but also displace as the door is opened and closed many times. An almost invisible gap between the gasket and the mating surface – for example, one that is an inch or two long and just a few hundredths of an inch wide – is sufficient to cause a substantial drop in shielding effectiveness (as much as 20–30 dB).

RF cables penetrating the chamber walls are an obvious conduit for unwanted interference; external signals can be picked up and conducted into the chamber on the outside of the cable shield, or even by the center conductor. Fortunately, good-quality RF cables and connectors are widely available and offer considerable protection against external interference. Typical shielding effectiveness of properly installed connectorized double-shielded coaxial cables can be 95–110 dB or more, and is sufficient to prevent interference.

The power and network cables that may be required to support the DUT or the test equipment is another matter altogether. Any metallic conductor entering the chamber from outside acts as an antenna, picking up and transporting external RF energy into the chamber. It is usually impractical to fully shield these cables; even if they can be shielded, there is no guarantee that the equipment (such as power supplies) to which they are connected are immune to RF pickup. Instead, filters are used at the points where the cables penetrate the chamber wall. Typically, L-C low-pass filters are used, with a cutoff frequency that is well below the frequency band or bands of interest. The filters should provide least 50–60 dB of attenuation at these frequencies.

In some situations, it may not be possible to filter out external interference without also removing the desired signals that must travel over the cable. For example, a Gigabit Ethernet network cable must carry signal bandwidths in excess of 100 MHz. A filter that can adequately suppress stray signals in the 2.4 GHz band while still presenting low insertion loss and passband ripple in the 0–100 MHz frequency range is not simple to design. In these cases, it is usually preferable to use an optical fiber cable (with the appropriate converters) instead of metallic conductors.

3.5 Conducted Test Setups

As previously mentioned, conducted test setups are simple, compact, and should be used if at all possible for any measurement that does not involve the DUT's antenna patterns. This is by far the most common test setup used in laboratories and manufacturing lines.

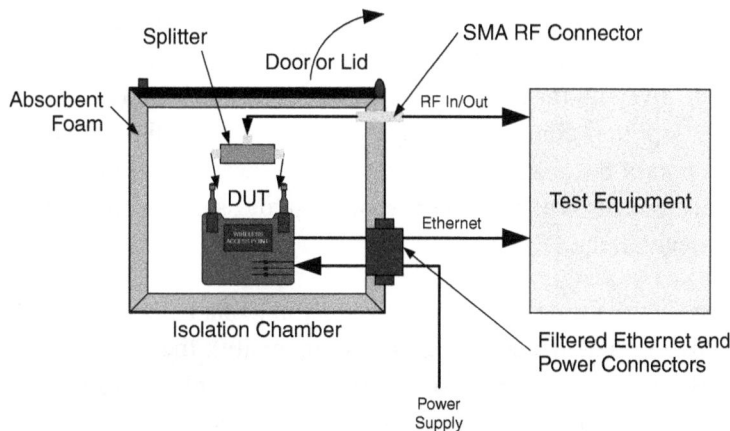

Figure 3.9: Typical Cabled Test Set up

If the DUT's antennas are connectorized and thus removable, and its case or enclosure is all-metal (and adequately shields the internal circuitry), no chamber is required and direct cable connections are possible. Otherwise, the DUT should be placed within an RF-tight shielded enclosure to protect the test setup from external interference. The enclosure is usually quite similar to the anechoic chamber described previously, but much smaller, because near-field/far-field issues do not apply here.

3.5.1 Coupling to the DUT

Coupling the RF signals to the DUT is best done by simply disconnecting the antenna(s) and substituting RF cables terminating in the appropriate connectors. Typically an adapter may have to be used. This is particularly true for commercial WLAN equipment. The Federal Communications Commission (FCC) requires "reverse-polarity" connectors (reverse-SMA, reverse-TNC, etc.) to be used for antenna jacks to prevent consumers from attaching high-gain antennas and amplifiers. Adapters are thus needed for connection of normal SMA cables during laboratory testing.

If the DUT uses internal (built-in) antennas, it is not usually possible to directly connect cables to it. In this case, a near-field probe (see above) is used to couple signals to and from the DUT. The near-field probe should be placed very close to the DUT's antennas, to ensure that maximum coupling is obtained, and to exclude as much of the DUT's self-generated noise as possible.

In some situations (particularly during development) it may be possible to open up the DUT's enclosure and terminate RF cables directly at the antenna ports of the DUT. This should only be done if it can be ensured that the cables, and the method of connecting them, do not result in a mismatch.

Most WLAN APs and many clients utilize diversity antennas, and hence there will be two antenna jacks on the DUT. If the diversity performance of the DUT is not being tested, it is better to remove the diversity function as a factor entirely. The best way of doing this is to use a 2:1 power divider (splitter) connected to the two antenna ports; the test equipment can then drive the common port of the splitter. In this case the same signal will be seen on both antenna ports, and the measurements will remain the same regardless of which one is selected by the diversity algorithm within the DUT.

An alternative means of working around the diversity antenna issue is to manually configure the DUT to use only one antenna port (i.e., turn off diversity); the other port should then be terminated. If all else fails, simply drive one of the diversity antenna jacks and terminate the other one; in most cases the DUT's diversity algorithm will select the driven jack for reception and ignore the terminated jack.

3.5.2 Power Levels

It is essential to ensure that power levels in conducted setups are well matched to the signal levels tolerated by both the DUT and the test equipment. Failure to do this leads to all manner of anomalous results, ranging from an unusually high bit error rate to permanent equipment damage.

Both the DUT and the test equipment typically contain sensitive radio receivers; as with any receiver, the signal levels must be matched to the dynamic ranges of the equipment. A signal that is too weak will be received with errors, or not at all; a signal that is too strong causes clipping or even intermodulation distortion, which also causes errors. In the case of sensitive equipment such as power meters, the full output of a WLAN AP – which may exceed 20 dBm, or 100 mW – can damage the power sensor, or cause it to go out of calibration. Thus the signal levels at all points in the test setup must be checked and adjusted before running a test.

The best method of doing this involves placing calibrated splitters or directional couplers at the RF inputs to the DUT as well as the test equipment, and then using RF power meters to determine the signal levels. Either fixed or variable attenuators can be used to reduce the signal levels to acceptable limits. If power meters are not available or usable, the RSSI of the DUT itself can be used as a rough indicator of whether the DUT is being overloaded.

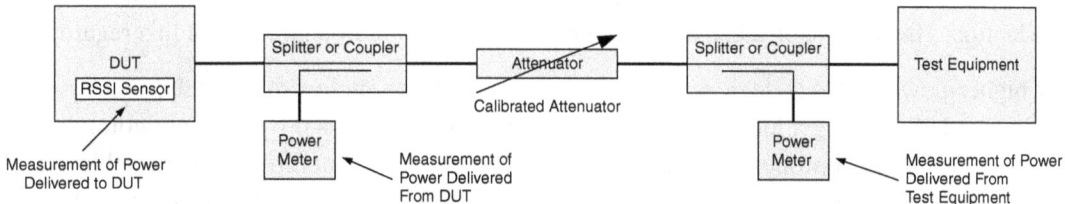

Figure 3.10: Power Control Methods

The signal strength input to the DUT should usually be placed in the middle of the receiver dynamic range, as this normally represents the best compromise between adequate signal-to-noise ratio (SNR) and intermodulation distortion. Note that the bottom of the dynamic range depends on the modulation format being used: complex modulation formats such as 64-QAM Orthogonal Frequency Division Multiplexing (OFDM) require SNRs in excess of 27 dB to achieve a tolerable bit error ratio, while simple formats such as Binary Phase Shift Keying (BPSK) require only about 3–5 dB SNR.

3.5.3 Excluding Interference

Excluding interference during conducted tests is simpler than for the other test setups. For one, the equipment is usually all placed close together and connected with relatively short cables. Further, it is relatively straightforward to ensure good isolation. The techniques described for chambered tests are applicable here as well.

One issue that is often overlooked is the need to ensure that *all* devices in the RF path provide adequate isolation. Good-quality test equipment poses little problem in this regard, as such equipment is almost always designed to generate very little electromagnetic interference and also provide extremely good rejection of external signals. However, unexpected sources of leakage are: unterminated ports on power dividers and directional couplers, short pieces of low-isolation cables, loose or improperly torqued connectors, and so on.

Of course, there is also no substitute for good old-fashioned common sense in this regard; for instance, it is not uncommon to find people (particularly software developers) carefully cabling up a test system with high-quality cables and enclosures, and then cracking open the door of the enclosure to connect an RS-232 console cable to the DUT!

3.5.4 Heat Dissipation

One often underestimated problem is that of dissipating the heat from a wireless device that is placed in an enclosure. An RF-tight enclosure does not allow free movement of air, unless specially manufactured vents are included.

For low-power WLAN devices (10 W dissipation or less), it is sufficient to provide enough air volume to conduct heat to the enclosure walls by convection, and then to the outside air by conduction. Having one or more of the walls be bare or painted metal helps in this regard.

For higher-power WLAN devices, an airflow path within the enclosure is highly recommended. Drilling a lot of vent holes in the enclosure walls is obviously not going to work, as the isolation properties will be destroyed before sufficient airflow can be achieved. Instead, air vents in enclosures are covered by means of thick sheets of honeycomb grille material. The length to width ratio of the grille apertures are large enough to cause the waveguide cutoff frequency of the grille to be quite high; the grille thus provides a high RF attenuation without restricting airflow. A small "muffin" fan can be placed in front of the grille to further enhance airflow.

3.6 Repeatability

Measurement repeatability is significantly affected by the type of environment chosen to conduct wireless testing, as well as the usual factors in any type of network testing. Repeatability may be a function of time (i.e., how well do the results of the same measurement performed at different times correlate?) or a function of space (i.e., if the same measurement is performed at different physical locations – for example by different laboratories – how well do the results correlate?). It is essential to strive for as much repeatability in both areas as possible; "one-time" measurements have very little value to anybody.

As may be expected, conducted environments provide the highest level of repeatability for any type of measurements. As all sources of external interference and variations in propagation behavior have been minimized, the remaining source of variation (besides random thermal noise), from one measurement to the next, reside in the test equipment itself. With good test equipment, this can be made very small. Further, it is possible to easily replicate a measurement, because the entire test setup can be accurately characterized, described, and reproduced elsewhere.

Well-constructed chambered OTA environments are close to conducted environments in terms of repeatability. The primary sources of variation here are in the ancillary equipment (jigs, fixtures, etc.), the arrangement of cables within the chamber, and the incidental emissions from the equipment present in the chamber along with the DUT. All of these can be minimized; for example, cable runs can be carefully positioned and described so that they can be reproduced if necessary.

As noted previously, indoor environments pose significant challenges for repeatable measurements. The nature and level of external interference is often completely beyond the control of the person carrying out the test, and can vary significantly from one measurement to the next. Further, the propagation characteristics change by very large amounts with even small changes in position or orientation within an indoor environment, making it difficult to set up and repeat measurements over time. Finally, each indoor environment is quite different from the next; thus it is virtually impossible to repeat measurements in different buildings.

Some degree of repeatability within indoor environments may still be achieved by means of statistical methods, however. Recent work by Airgain Inc. indicates that, while any one measurement at a single location bears little correlation to the next, the average of a large number of measurements at random locations and orientations can be reasonably repeatable.

The environment is not the only source of problem when trying to repeat a measurement, however. The source of test stimulus must also be considered carefully. For example, many WLAN performance test setups – especially those at the system level – try to use standard PCs or laptops as software traffic sources. The traffic generated by these devices shows wide variations over time (due to interactions between the operating system and applications), and is also dependent on the precise collection of software and device drivers loaded on the computer. Laptops were never designed to be test equipment, and do not lend themselves to accurate generation of WLAN traffic.

Physical Layer Measurements

Measurements and test methods applicable to the WLAN Physical (PHY) layer, principally 802.11a/b/g PHYs, are dealt with in this chapter. (The emerging area of MIMO PHYs (i.e., 802.11n) is covered in Chapter 10.) Also, manufacturing test, which involves a large number of PHY layer measurements, is dealt with in Chapter 7.

4.1 Types of PHY Layer Measurements

PHY layer measurements are performed for various reasons:

- functional tests done during design and development of wireless LAN (WLAN) equipment and chips;

- performance tests, which try to quantify how well the RF and baseband portions of the equipment works;

- characterization tests used to determine how the equipment or chipset performs under various conditions;

- compliance measurements, done to ascertain whether the equipment conforms to government-mandated emissions requirements, or to specifications in the 802.11 PHY standard.

This chapter will focus on the tests and measurements that are performed on completed devices or systems. Measurements on chipsets are carried out on a reference implementation, such as the chipset vendor's reference design or evaluation board. Measurements on box-level products are done using prototypes, or even the final production version. It is extremely unusual to make PHY layer measurements at the large-system level (i.e., involving more than one Access Point (AP) or client at a time).

4.1.1 Design and Development

A variety of PHY layer measurements are performed by engineers during the design and development of WLAN equipment (or chipsets). By far the widest range of RF test instruments is used here: network analyzers, spectrum analyzers, oscilloscopes, signal generators, power meters, noise figure meters, frequency counters, and logic analyzers, to name but a few. Many books have been written on RF and digital signal measurements during the development process, and their contents will not be repeated here.

4.1.2 Transmitter Performance

Transmitter performance tests are aimed at quantifying how well the transmitter functions of a WLAN device (typically comprising baseband digital signal processing, up-conversion, filtering, and power amplification) are implemented. Besides the obvious test of power output, transmitter tests also measure the quality of the output signal, typically using a measurement such as error vector magnitude (EVM). Additional tests performed include channel center frequency accuracy, range over which the output power can be controlled, the delays incurred when turning on or off the transmitter, and the degree of matching to the antenna (VSWR).

4.1.3 Receiver Performance

Receiver performance tests are likewise aimed at quantifying the capabilities of the receiver functions: front-end amplification, down-conversion and filtering, and baseband digital signal processing. Sensitivity and dynamic range are obvious candidates; others include rejection of co-channel and adjacent channel interfering signals, the ability of the receiver to detect a valid signal in its channel (clear channel assessment, or CCA), the time taken to lock to an incoming signal, and the accuracy of the received signal strength indication (RSSI).

4.1.4 Device Characterization

Characterization tests are done to determine how well the device performs over environmental parameters, such as voltage and temperature. In the case of chips, this is done at various process 'corners' as well. Usually, characterization tests involve repeating a key subset of the performance tests at different supply voltages or ambient temperatures, and verifying that the device continues to meet expectations in these areas at the rated extremes of voltage and temperature.

4.1.5 Compliance

Wireless devices are subject to a much wider variety of regulatory requirements than their wired brethren; in fact, critical factors such as permissible output power, spurious emissions limits, and usable channels may even vary from country to country. Extensive compliance tests are therefore needed to verify that the device is legally capable of being sold for use in specific areas of the world. Compliance tests cover radiated power, spurious emissions, distribution of power in the transmitted spectrum (spectral mask compliance), channel occupancy, and avoidance of interference to primary spectrum users (e.g., radar detection and avoidance).

4.2 Transmitter Tests

Transmitter tests, during both general design and development as well as for performance testing during design verification, use the following basic setup. Note that there are many variations depending on the exact nature of the test being conducted, but the key pieces of equipment remain more or less the same. The figure below represents this setup in schematic form.

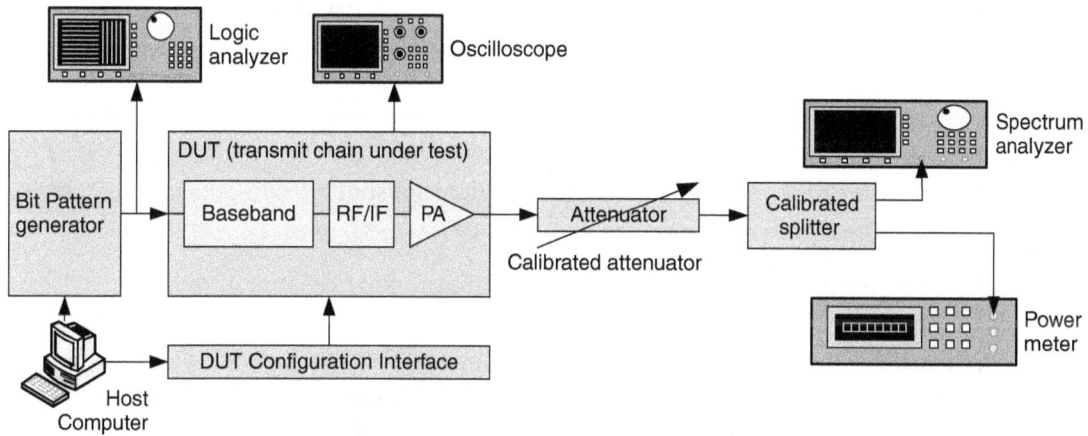

Figure 4.1: General Transmitter Power/Frequency Test Setup

Design engineers almost always construct, test, and optimize each piece of the transmitter chain separately, as it is far easier to find bugs in this manner. However, once the entire transmit chain (baseband, upconverter, and power amplifier (PA)) has been individually verified to function, the pieces are put together into a single module and tested as a unit to verify that performance meets expectations and datasheet specifications. This section deals with such performance tests.

With reference to the figure above, the various elements of the test setup (besides the Device Under Test (DUT)) are:

• A bit pattern generator or some other means of driving the baseband device with digital data to be transmitted. In some cases this is accomplished using a PC with a parallel I/O interface, or even a test mode in the actual medium access control (MAC) device that will eventually be used with the transmitter. See a subsequent section in this chapter for methods of forcing the transmitter to generate output data.

• A logic analyzer to verify the data that is actually driven.

- A calibrated attenuator for reducing the output signal power from the DUT's PA before driving sensitive devices such as power meters. A typical WLAN PA may have a power output of $+20\,dBm$ (100 mW) which can damage or destroy power meters and some VSAs.

- A spectrum analyzer (more usually a VSA) for power spectrum and linearity measurements, as well as gross power measurements.

- A power meter for more accurate measurements of RMS output power.

- An oscilloscope for various purposes, such as measuring turn-on and turn-off delays (though some VSAs also offer functions for these purposes).

In addition to the above instruments, frequency counters (not shown) may be employed to measure the frequency accuracy and stability of the local oscillators and synthesizers. Note that modern high-end VSAs are often accurate enough to substitute for frequency counters, especially with high-stability reference oscillators fitted.

Triggering of spectrum analyzers and VSAs is a significant issue because of the wide variations in power of 802.11 signals within a frame (e.g., between the preamble and the data portion, and even within the data portion of orthogonal frequency division multiplexing (OFDM) frames) as well as the irregular intervals at which frames are transmitted. Fortunately modern VSAs are equipped with advanced trigger capabilities that, if carefully adjusted, solve this issue quite well. In addition, they also offer trigger hold-off capabilities so that they can avoid retriggering, for example, on the acknowledgement frame that immediately follows a data frame of interest.

4.2.1 Output Power and Power Control Range

The output power test measures the RMS output power averaged over the entire occupied channel. Accurate power measurement of packetized signals with complex modulation such as OFDM or CCK is a problem; traditional average or peak power measurement tests will not work in this case. This test is usually performed with an accurate time-gated power meter during characterization of the transmitter module. The module generates frames of a specified duration, and the power meter measures the RMS power averaged over the duration of the frame. The trigger capabilities of the power meter are used to isolate the power measurement interval to the duration of the frame; a power threshold setting is established for the trigger point, and then the power measurement interval is set starting from the trigger point. Thus the power meter will only measure power over the duration of the frame.

If the RF chain can be made to generate frames at a fixed rate, then it is possible to perform power measurements with a standard averaging power meter, by measuring or calculating the mark/space (on/off) ratio of frame transmission time to idle time and then adjusting the measured average by the reciprocal of the mark/space ratio. This is, however, a less accurate and straightforward approach.

A vector signal analyzer (VSA) is a more convenient instrument for measuring power, albeit less accurate than a power meter. The advanced triggering and arithmetic capabilities of the VSA simplify the measurement process. The VSA is most commonly used on the lab bench for quick measurements of transmit power.

Besides absolute power measurements, it is necessary to measure and characterize the power control range of the transmit chain. WLAN systems are expected to allow user-controllable power settings, usually ranging from around 0 dBm average output power to the maximum output level (usually between $+15$ and $+23$ dBm). Operation in the 5 GHz range is particularly sensitive in this regard; transmit power control (TPC) facilities are mandated by regulatory bodies in order to limit interference with other licensed services, such as radars. For this reason, the power control range and power setting accuracy of the WLAN transmitter need to be measured and calibrated to match the datasheet specifications for all operating channels and PHY bit rates. The measurement procedure is as described above, except that the output power is stepped through its range and a series of measurements made.

4.2.2 Noise, Linearity, and Distortion

These tests measure the general quality of the transmitted signal and expose issues in the components and design of the transmit datapath. They are usually done on the RF chain prior to integration with the baseband. They generally assess the quality of the local oscillator (i.e., the frequency synthesizer that supplies the LO signal for upconversion), and the dynamic range and linearity of the RF/IF chain and the PA.

Linearity and gain/phase flatness measurements are usually done with a network analyzer, both stage by stage as well as on the whole chain. The setup in Figure 2.4 is normally used. This measurement can also expose stability problems in the chain, particularly at frequencies outside the normal operating bandwidth of the system, by indicating where gain peaks and rapid phase changes occur. It is essential therefore to sweep over a much wider band of frequencies than the usual operating zone of the RF/IF chain.

Noise and distortion measurements are performed by driving the RF/IF chain with a signal generator, usually a vector signal generator (VSG), and then using a VSA to look at the output. The VSG can provide actual OFDM or CCK signals, enabling for instance measurements of 'spectral regrowth' due to PA nonlinearities.

The spectrum analyzer can show the phase noise characteristics of the output signal when a continuous wave (CW) signal generator is used to drive the input. For really accurate measurements on the LO and synthesizer, a phase noise test set can be used, but this is usually overkill in the case of WLAN devices.

The 802.11a/g PHY standard also specifies spectral flatness in terms of the variation of the average energy of the subcarriers (the variation must be less than $+/-2$ dB for the center 32 subcarriers, and $+2/-4$ dB for the outer subcarriers). This is easily measured using a VSA, as most VSAs can display the averaged energy of each individual subcarrier as measured over an OFDM frame (or over multiple OFDM frames).

4.2.3 EVM

EVM is a metric of modulation quality that has become a standard measurement in a number of different digital radio technologies (cellular, cable television, WLANs), particularly those using some variant of QAM (Quadrature Amplitude Modulation). It is generally used to assess the transmitter signal quality and determine the degree of signal impairment (i.e., deviation from the ideal expected signal). EVM has become an important figure of merit in WLAN transmitter performance tests. The following figure represents the EVM concept.

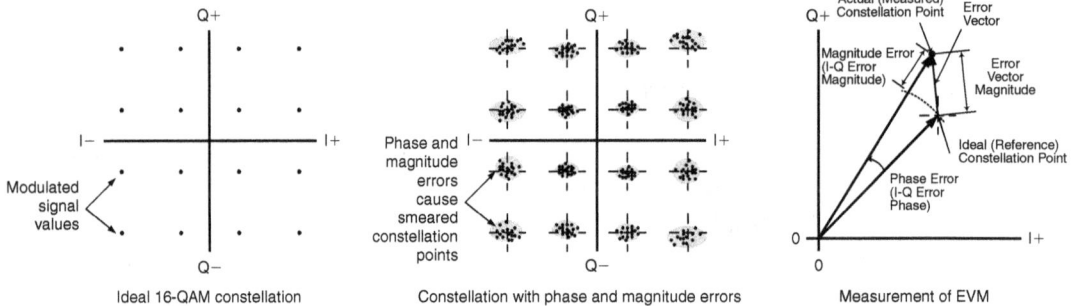

Figure 4.2: EVM

The principle underlying the EVM measurement is as follows. QAM modulation consists of a series of complex numbers that are used to modulate a high-frequency carrier by altering its amplitude and phase. The ideal modulated transmit signal can be represented as a set of *constellation points* on a 2-D complex plane, with the real and imaginary values of the signal falling along the X and Y axes, respectively. Thus the 64-QAM signal used for 48 and 54 Mb/s OFDM modulation in 802.11a/g has 64 constellation points. Then, as shown in the figure above, the errors in the transmitted signal can be represented as vector deviations from these points. Essentially, the impaired transmitted signal at any time instant can be considered as the vector sum of an ideal signal and an error signal.

Unfortunately, neither the ideal signal nor the error signal can be measured directly; they occur within the transmitter chain itself (i.e., within the DUT) and are not accessible to external equipment. Instead, the EVM measurement reconstructs the digital data by demodulating the transmitted signal, maps the digital data to the ideal constellation points, and then subtracts (vectorially) these ideal points from the actual measured constellation points. (This is represented in Figure 4.3.) The process of subtraction yields the error that was introduced during the modulation process by the DUT. This effectively means that measuring EVM requires a receiver function to be implemented first (to determine what data was transmitted), which in turn assumes that the receiver output is error free. As bit errors are always present to some degree in every practical receiver, the EVM measurement process is not perfect.

The error vectors calculated for each modulation symbol from the EVM measurement process are then averaged on an RMS basis over some defined length of data (e.g., 1000 symbols, or even an entire frame) and then expressed as a percentage; this is the EVM. Note that only the absolute magnitude of the error vectors are taken, the phase of the error vector is not relevant to the EVM measurement. A 100% EVM indicates that all of the constellation points of the transmitted signal have shifted or 'smeared' into other constellation points, and the original signal cannot be successfully recovered because no constellation point can be distinguished from another. (In fact, complex modulation schemes such as 64-QAM are completely unrecoverable with as little as 35% EVM.) On the other hand, a 0% EVM indicates that the transmitted signal exactly matches the ideal, which is, of course, impossible. (The EVM is therefore also equivalent to the relative constellation error of the modulated signal.)

The error signals measured and quantified by the EVM metric encompass all of the sources of error that are introduced during the modulation and transmission process, including:

- random noise (both amplitude and phase) introduced by amplifiers, mixers, and oscillators;
- conversion errors in the D/A converters;
- distortion throughout the RF chain;
- filter imperfections such as passband ripple or nonlinear group delay;
- mixer spurious signals;
- digital signal processing errors;
- offsets and imbalance in the quadrature mixers.

The EVM measurement thus assesses the quality of the transmitted signal with the entire baseband/IF/RF chain in place. It provides an excellent overall picture of the transmitter quality, because it factors in everything (phase noise, nonlinearity, etc.) and essentially reflects the level of difficulty that a conforming receiver will have when trying to demodulate the transmitted signal. Excessive EVM numbers immediately point up issues in the transmitter that must be fixed. Further, direct inspection of the smeared constellation can often quickly indicate the source of the underlying problem, such as I/Q imbalance.

In the case of 802.11 WLANs, the requirement is that the EVM be measured as an average over 20 complete data frames, each of which should be at least 16 symbols in size, and containing random data. The IEEE 802.11 standard actually defines a procedure for measuring EVM for 802.11a and 802.11g transmitters (see subclause 17.3.9.7 of IEEE 802.11), which in theory can be done with a simple setup consisting of a mixer, a stable microwave signal generator, and a digitizing oscilloscope. The process as described is roughly:

a. Mix down the transmitted signal to baseband and use the sampling oscilloscope to take complex-valued (i.e., amplitude and phase) samples at a sampling rate of at least 20 megasamples per second or more.

b. Capture these samples into a signal processing function of some kind (e.g., a software program running on a computer) and decompose them into in-phase (I) and quadrature (Q) components.

c. Implement an approximation of the OFDM receiver in the signal processing function, including preamble detection, frequency offset compensation, extraction of pilot subcarrier values, and normalization and demodulation of the subcarriers, producing one constellation point per OFDM subcarrier.

d) For each of the 48 data-carrying subcarriers, find the closest 'ideal' constellation point and calculate the absolute (Euclidean) distance to the actual constellation point. As the constellation points are represented in terms of I and Q components, the RMS average is used to simultaneously generate the Euclidean distance and take the average of all the points in the frame.

The following equation (Equation 33 from Clause 17 of IEEE Std 802.11-2007) is used to calculate the EVM once the I and Q values of the constellation points of all the subcarriers and all the symbols have been found:

$$
\text{Error}_{\text{RMS}} = \frac{\sum\limits_{i=1}^{N_j} \sqrt{\dfrac{\sum\limits_{j=1}^{L_p} \left[\sum\limits_{k=1}^{52} \{(I(i,j,k) - I_0(i,j,k))^2 + (Q(i,j,k) - Q_0(i,j,k))^2\} \right]}{52 L_p \times P_0}}}{N_f}
$$

where

L_P is the length of the packet;

N_f is the number of frames for the measurement;

$I_0(i, j, k)$ and $Q_0(i, j, k)$ denotes the ideal symbol point of the ith frame, jth OFDM symbol of the frame, kth subcarrier of the OFDM symbol in the complex plane;

$I(i, j, k)$ and $Q(i, j, k)$ denotes the observed point of the ith frame, jth OFDM symbol of the frame, kth subcarrier of the OFDM symbol in the complex plane;

P_0 is the average power of the constellation.

Note that the EVM is actually specified in the 802.11 standard in terms of a relative constellation error (in dB relative to the average power of the constellation). It may be easily converted into a percentage relative to the average amplitude of the symbols, as follows:

$$EVM_{\%RMS} = 10^{(EVM_{dB}/20)}$$

The following table gives the maximum permissible EVM for various OFDM (802.11a/g) modulation formats.

OFDM data rate (Mb/s)	Modulation type	Constellation points	EVM (dB)	EVM (%RMS)
6	BPSK	2	−5	56.2
9	BPSK	2	−8	39.8
12	QPSK	4	−10	31.6
18	QPSK	4	−13	22.3
24	16-QAM	16	−16	15.8
36	16-QAM	16	−19	11.2
48	64-QAM	64	−22	7.9
54	64-QAM	64	−25	5.6

In practice, measuring EVM on 802.11 devices is almost always done using a demodulating spectrum analyzer (i.e., a VSA), as this is by far the most convenient and accurate method of doing so. (At 54 Mb/s, the allowable EVM is only 5.6%, implying that the measurement system itself has to be highly stable and linear to permit accurate measurements.) In addition, of course, some means has to be found of driving the DUT baseband with sufficiently long frames and of the right PHY data rate. The typical EVM test setup is shown below.

The test setup is actually relatively simple because the VSA does all the work, ranging from demodulating the transmitted signal to calculating the EVM and even providing triggers so

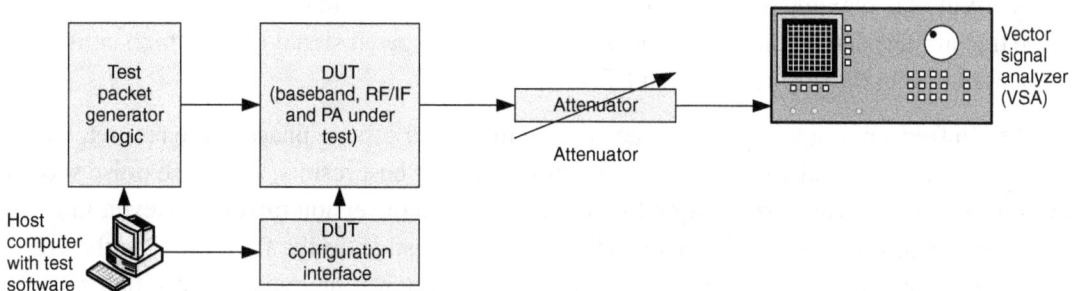

Figure 4.3: EVM Test Setup

83

that measurements can be made on the correct frames (i.e., the data frames transmitted by the DUT). The frame source drives the baseband with digital data in the form of complete PLCP-formatted frames. The VSA locks to the transmitted frames, demodulates them, and measures the EVM; in a sense, therefore, the VSA acts as an 'ideal' receiver. The equalization function of the VSA should be turned on to remove frequency error as much as possible before measurement.

Modern VSAs have very powerful software packages, specialized for different wireless technologies (including 802.11a/g) that can implement advanced triggering and filtering techniques to simplify the measurement. Some VSAs (such as the ESA series from Agilent Technologies) also include the ability to display EVM separately for each subcarrier in the OFDM signal, which can point up issues in the filters or the baseband itself. VSAs can also display EVM variations over time within a single captured frame, which is useful for detecting LO stability errors and other artifacts within the transmit datapath.

4.2.4 Frequency Stability

Frequency stability refers to the ability of the carrier oscillator in the DUT to provide a CW signal that corresponds as closely as possible to the nominal channel center frequency, as measured over different time scales. Frequency accuracy can be generally described by two parameters: short-term frequency stability, which is also termed as phase noise or period jitter, and long-term frequency stability, which is generally referred to as aging. (For more information, consult IEEE Std 1139-1988, 'Standard Definitions of Physical Quantities for Fundamental Frequency and Time Metrology'.)

Short-term frequency stability is the most important in the case of WLANs, as it directly affects parameters such as EVM and receiver performance. The impact of phase noise on EVM is quite simple to understand: a carrier oscillator that has high phase noise will impose this phase noise as a phase error on the modulated signal, which when demodulated will produce constellation points that are smeared over a large region. The phase noise effect is the same regardless of whether the noisy oscillator is in the transmitter or in the receiver. The 802.11 PHY specifications do not discuss phase noise directly, but instead indirectly limit it by specifying the maximum transmit EVM (which places constraints on the upconversion LO in the transmitter) and the maximum packet error rate at a given signal level (which constrains the phase noise in the downconversion LO in the receiver).

Short-term frequency stability is most accurately measured using a phase-noise test set, which is a rather expensive and complex piece of equipment. For best results, the phase noise should be measured at the synthesizer output to the receive downconversion mixer. However, in integrated solutions, this signal is not usually externally visible, as the frequency synthesizer is typically integrated on the same chip as the downconverter. If the reference oscillator input to the synthesizer is available, however, an approximation to the lower limit of the actual synthesizer phase noise may be obtained by measuring the phase noise of the reference and then using the formula:

$$\phi_{synth} = \phi_{ref} \times 20\ log_{10}(F_{synth}/F_{ref})$$

This is possible because the close-in phase noise of the synthesizer is essentially governed by the reference, and the synthesizer multiplies the phase noise of the reference by the logarithm of the upconversion frequency ratio. The above does not take into account the noise introduced during the frequency synthesis process or the VCO noise. Unfortunately many WLAN chipsets use integrated CMOS or BiCMOS VCOs, which are notoriously noisy oscillators.

An alternative, if a phase-noise test set is not available, or the VCO noise cannot be ignored, is to drive the transmit chain (after the baseband) with a clean, high-stability signal generator and use a high-dynamic-range spectrum analyzer to measure the phase noise of the output signal. As already noted, integrated VCOs are quite noisy – often greatly exceeding the noise floor of a high-end spectrum analyzer. This permits a reasonable phase noise measurement to be made by setting the transmit chain to generate a CW signal, as described in Section 4.6.2, and directly inspecting the signal with a spectrum analyzer set to a low resolution bandwidth (RBW). (An even simpler alternative is to simply perform an EVM test, and assume that if the EVM is low the phase noise must be likewise low.)

Initial accuracy and long-term stability (drift) are principally controlled by the reference source, usually a crystal oscillator, that is used in the frequency synthesizer. Quartz crystals change frequency with age as a consequence of mechanical changes in their mounting structures (stress relief), contamination of the crystal surfaces, or parametric changes in the surrounding oscillator circuitry; frequency changes with time may range from 1 to 10 parts per million per year. In addition, temperature changes (e.g., as the WLAN DUT warms up during operation) causes frequency drift, which can be 5 to as much as 100 parts per million over the operating temperature range of the DUT. Finally, the crystal oscillator may differ from the nominal (vendor specified) frequency by anywhere from 5 to 50 parts per million, depending on the manufacturing tolerances of the device. All of these contributing factors cause the actual reference frequency supplied to the LO synthesizer to differ from the nominal (expected) frequency.

Aging and drift are of somewhat less importance than phase noise, as the demodulators perform a carrier extraction, locking and tuning process for each received frame in order to compensate for frequency differences between the transmit synthesizers of the various WLAN clients and APs that may all be part of the same basic service set (BSS). However, one requirement that must be verified is that the initial accuracy and the drift should be within the 802.11 specifications for the PHY under use. The limits are +/−25 ppm for 802.11b and +/−20 ppm for 802.11a/g, which is fairly tight and mandates a high-quality crystal oscillator, sometimes even a TCXO (temperature compensated crystal oscillator).

Verification of drift is done by configuring the transmitter to output a CW signal (i.e., using a test mode as described below) and using a high-stability frequency counter to monitor the output over time, temperature, and supply voltage changes. Care should of course be taken to derive the frequency counter reference from a source that has higher stability than the WLAN crystal oscillator; the internal reference of most general-purpose frequency counters is usually a TCXO that may not have much better stability than the WLAN DUT itself.

4.2.5 Turn-On and Turn-Off Ramp

RF transmitters cannot go from zero to maximum power (or vice versa) instantly; they take time to build up or bring down the output power level. The time required for the transmitter to settle to within some predefined tolerance (usually 10%) of the target power, either maximum or zero, is referred to as the turn-on or power-on ramp and the turn-off or power-off ramp, respectively.

The turn-on and turn-off ramps are limited to $2\,\mu s$ in the 802.11 standard for 802.11b PHYs; for OFDM PHYs it is not specified, but a transmit-to-receive and receive-to-transmit maximum turnaround time of $5\,\mu s$ is specified instead, which essentially limits the transmitter turn-on and turn-off ramps to substantially less than this value (i.e., to about $2\,\mu s$ as well). These ramp times are important performance parameters as they govern the speed at which an 802.11 device can respond to a frame transmitted to it by another device. Also, a long turn-on delay can eat into the OFDM training sequences transmitted at the beginning of the frame, leading to less accuracy of frequency offset estimation at the receiver and consequently an increased bit error rate (BER). Finally, excessive turn-on or turn-off delays or power oscillations during turn-on can point up problems with PA linearity and stability, frequency synthesizer stability, supply decoupling, and overall system slew rate. These delays are therefore quite important to measure, as well as any overshoot or ringing when turning on.

The most logical method of measuring these is to transmit a steady stream of frames and trigger an oscilloscope with the leading or trailing edge of the transmitted signal (for the turn-on and turn-off ramp, respectively). However, 2.4 and 5 GHz OFDM signals are usually beyond the capabilities of all but the most expensive oscilloscopes. Instead, this is usually measured with VSA, triggered from the signal and displaying the signal envelope. After the appropriate ramp is identified, markers and measurement cursors on the VSA can be used to measure the ramp time. Note that this can also be measured with an envelope detector and a scope, but this requires careful attention to system setup and calibration.

In Figure 4.4, a VSA is used to measure turn-on and turn-off time. By setting up triggers appropriately, the VSA can be configured to capture and display the envelope of the signal at the start or the end of the transmitted frame. If an external device such as a standard client or AP is required in order to persuade the DUT to transmit data frames (see Section 4.6), then care should be taken to avoid measuring the envelope of the external device instead of the DUT

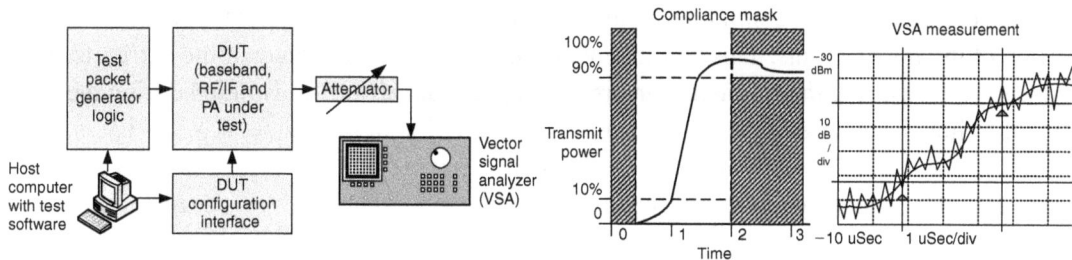

Figure 4.4: Turn-On and Turn-Off Time Test Setup

4.2.6 VSWR/Return Loss

At HF/VHF, VSWR is measured directly using a VSWR bridge or return loss bridge (RLB), and employing a standard signal generator or the transmit PA itself as the signal source. However, at the microwave frequencies that are used by WLANs, VSWR is instead measured indirectly, using a network analyzer to measure the appropriate scattering parameters (S_{11} and S_{22}) of various elements in the path to the transmit antenna, such as attenuators, couplers, switches, and the antenna itself, and then calculating or modeling the VSWR that would result when driven by the actual PA. Matching is performed by measuring the S_{22} of the PA and then adjusting matching sections until the desired match is obtained.

4.2.7 Channel Center Frequency Tolerance

The transmitted channel center frequency error is an important parameter to measure and verify, because an excessive center frequency error can result in conforming receivers being unable to demodulate the transmitted signal, as their tuning and frequency offset compensation range may not be sufficient. The 802.11 standard specifies that the center frequency tolerance must be $+/-25$ ppm in the 2.4 GHz band and $+/-20$ ppm in the 5 GHz band. The center frequency must be verified for every operating channel.

A simple spectrum analyzer measurement of the transmitted spectrum of the DUT is insufficiently precise to allow measurement to the above tolerances. This is most accurately measured by employing a chipset test mode to cause the transmit datapath to output a steady (CW) carrier, and then using a frequency counter to determine the actual center frequency. (See Section 4.2.4 above on the measurement of frequency stability.)

If no test mode is available (or accessible) in the chipset, and an 802.11a/g (i.e., OFDM) PHY is being tested, then a particular requirement of the 802.11 standard may be employed in order to use a VSA for this purpose. Specifically, the 802.11 standard requires that the symbol clock used for generating and modulating the carrier, and the transmit center frequency reference, must be derived from the same source (i.e., phase locked loop (PLL) or crystal oscillator). This is necessary for ensuring that once the receiver is locked to the incoming preamble training sequences, it will continue to remain frequency and phase coherent when the data portion of the frame begins. This feature enables the frequency tolerance of the carrier to be deduced by measuring the frequency errors of the pilot subcarriers transmitted during the OFDM training sequences as well as during the data portion of the frame. A VSA that is capable of measuring and displaying these frequency errors can thus be used to determine the transmit center frequency tolerance as well.

The latter method does have an advantage, in that the center frequency errors can be measured with actual transmitted frames rather than simply generating a carrier. Further, oscillator drifts during the frame can be detected by this method as well (some VSAs even compute and display such drifts).

4.2.8 Total Radiated Power

Total Radiated Power, or TRP, is a basic system measurement. The TRP measurement integrates the radiated power from the DUT antenna in all three dimensions; effectively, it measures the RF power density produced by the DUT over the entire surface of an imaginary sphere with the DUT at its center. TRP is a valuable performance metric because it includes the effects of not only the active devices but also the DUT antenna, device enclosure, nearby cabling, etc. It thus produces a view of the complete device, which is useful for assessing the relative merits of two devices that have radically different antenna patterns. For example, if two DUTs with similar capabilities but slightly different antenna patterns are positioned identically in an open-air test setup, the measurement antenna may fall in a null of one of the DUT's radiation patterns and in the peak of the other DUT's radiation pattern; this produces a false impression that one DUT is inferior to the other. In fact, even cables and other metallic objects connected to or near the DUT can cause the radiation pattern to change radically, causing point measurements to be unreliable and not representative of the actual DUT performance. By integrating over a sphere, however, the TRP measurement 'integrates out' the radiation pattern and produces the true radiated power from the DUT.

The measurements required to calculate TRP can also be used to calculate a number of other useful DUT parameters, principally concerned with the DUT's PA and radiating system. For example, the peak EIRP (useful for regulatory compliance tests) is the highest point measurement made during the TRP test; the directivity is the ratio of the peak EIRP to the TRP; and the gain is the ratio of the peak EIRP to the transmit power actually output by the DUT PA. The efficiency of the DUT's radiating system can also be calculated as the TRP divided by the power output measured directly at the PA output terminals.

The following figure illustrates the test setup used for TRP measurements.

As shown, TRP measurements require the use of a large anechoic chamber and a 2-axis positioner. The anechoic chamber must provide a sufficiently large quiet

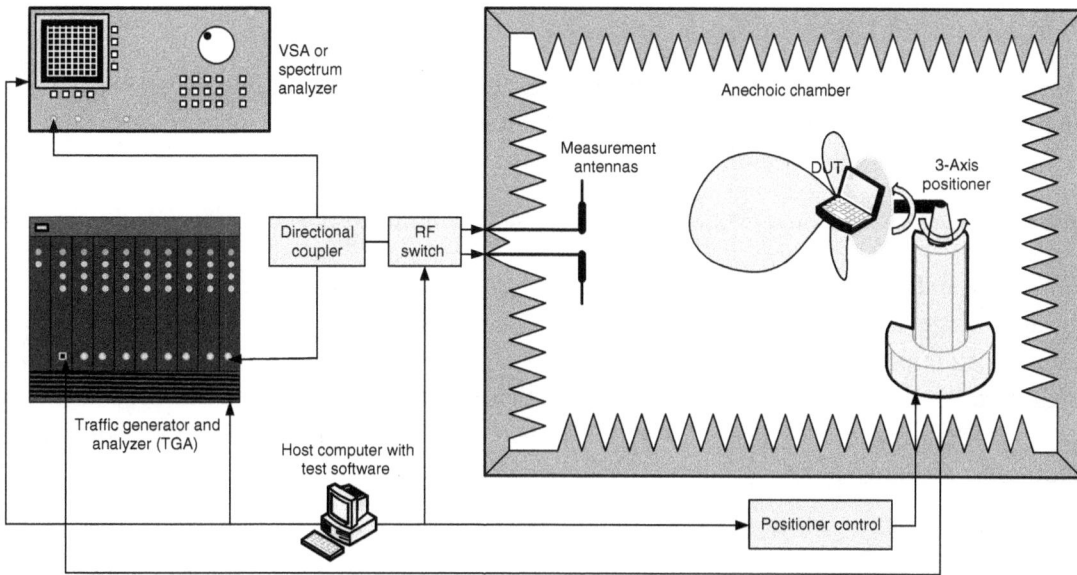

Figure 4.5: TRP Measurement Setup

zone within which the DUT can be rotated. The DUT is rotated in three dimensions, in steps of 5–15°, and a transmit power measurement taken at each point, thereby producing power measurements at all points of a grid superimposed on a spherical surface with radius equal to the distance between the DUT antenna and the measurement antenna. (This is similar to the intersection points of the lines of longitude and latitude that are superimposed on the Earth's surface.) A typical TRP measurement with 5° steps around the Z and X axes involves measuring transmitted power at each of $(360/5) \times (180/5) = 2592$ individual points. Once the radiated power P has been measured at all desired points surrounding the DUT by stepping the positioner over all angles θ_n, ϕ_m, it is integrated over the surface of a sphere to obtain the TRP. The equation used to calculate TRP from the individual measurements is:

$$T = \frac{\pi}{4N^2} \sum_{n=1}^{N-1} \sum_{m=0}^{2N-1} P(\theta_n, \phi_m) \cdot \sin(\Theta_n)$$

where T is the TRP, $\Theta_n = \theta_n \times \pi/180°$ and N is the number of steps between 0° and 180°.

A two-axis positioner is required in order to rotate the DUT in three dimensions within the anechoic chamber. Because of the large number of measurements to be taken and the number of different pieces of equipment that must be configured for each measurement, the TRP measurement is usually automated using a program or script running on the host computer. Even with an automated setup, these measurements can take several hours to complete for a single DUT configuration.

4.3 Receiver Tests

Various receiver performance and functional tests are also performed during general design and development as well as for design verification. Some of the more significant tests are described in this section.

4.3.1 Sensitivity, Dynamic Range, and Adjacent Channel Rejection

Sensitivity and dynamic range (blocking dynamic range, third-order intermodulation (IMD) dynamic range, etc.) are key parameters for any type of receiver. Adjacent channel rejection (ACR) is an important performance parameter for a channelized digital radio, which is what a WLAN receiver fundamentally is. For WLANs, these metrics are generally measured in terms of signal levels required to produce a BER less than or equal to some specified threshold, with a test setup as shown below:

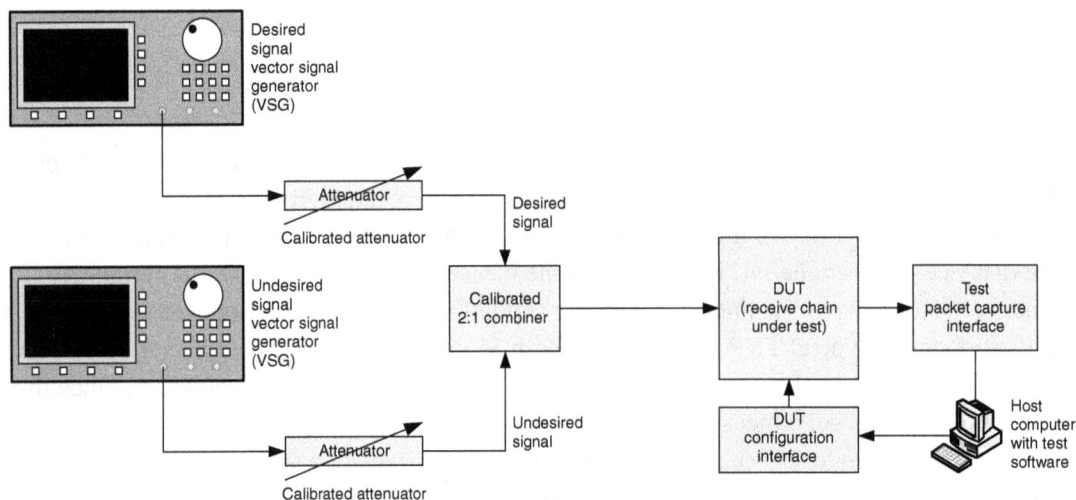

Figure 4.6: Sensitivity, Dynamic Range, and ACR Test Setup

The measurement of sensitivity is straightforward. Only one of the VSGs in the above setup is used; the other is deactivated or removed (and a terminator applied in its place). The active VSG is set to a fixed output power and configured to generate a stream of WLAN frames of a fixed frame size, with a fixed PHY bit rate, and at a fixed frame rate; the output power from the VSG is set to a convenient but known value. The attenuation is progressively increased until the received BER rises above a given threshold, depending on the 802.11 PHY technology used. The sensitivity of the receiver is then merely the VSG output level (in dB) minus the attenuation (also in dB).

As it is much simpler to measure frame error ratio (FER) (by using the frame check sequence (FCS) in each frame to detect a frame error) than BER, it is more common to substitute FER for BER. (BER can be translated to FER by simply multiplying by the number of bits in the frame; thus a BER of 1×10^{-5} for a 1000 byte frame corresponds to an FER of 8%). For 802.11b, the frame size used is 1024 bytes, and the FER level is set to 8%. For 802.11a/g, the frame size is 1000 bytes, and the FER level used is 10%. The actual sensitivity values are highly dependent on the specific receiver used -and in fact on the manufacturing tolerances for the receiver – but typical sensitivity figures range from −76 dBm to as good as −92 dBm for most modulation types. Generally, the lower the signal required to reach the FER threshold, the better.

Dynamic range is measured by finding the maximum input level (overload level) of the receiver, and then simply subtracting the sensitivity. This is basically the blocking dynamic range, and corresponds to the receiver maximum input level specification of the 802.11 standard. The measurement process used is similar, but instead of increasing the attenuation, the attenuation is decreased (and the VSG level potentially increased) until the received frames start showing bit errors. The same criteria in terms of FER is applied to the maximum input level, so that a consistent measurement of dynamic range can be obtained. As with sensitivity, the dynamic range is significantly affected by receiver design and manufacturing parameters. Typical maximum input levels for WLAN receivers range from −20 dBm to as high as 0 dBm, producing dynamic ranges in the region of 80–90 dB; the 802.11 standard requires a maximum input level tolerance of at least −30 dBm for 802.11a, −20 dBm for 802.11g, and −10 dBm for 802.11b.

Rather than the third-order IMD product measurements commonly used for analog receivers at HF and VHF, WLAN receivers are characterized in terms of ACR ratios. As the skirts of even a compliant 802.11 transmitter extend for a substantial distance on both sides of the center frequency – for example, an 802.11a or 802.11g transmitter may have signals as high as −40 dB relative to the in-channel power level at +/−30 MHz away from the center frequency – there is a significant need to reject adjacent channel energy in order to provide error-free reception in the presence of other APs and clients on nearby channels. The measurement method is quite similar to those performed for third-order IMD on analog receivers, and is performed after the receiver sensitivity is known. Two VSGs and attenuators are used, as shown in Figure 4.7; one VSG produces a signal in the desired channel (i.e., a signal that the receiver should demodulate successfully), while the other VSG is set to an immediately adjacent channel, either higher or lower than the desired-channel. The second VSG is basically used to interfere with the desired-channel signal.

The attenuators and VSG outputs are adjusted to produce a desired-channel signal that is 3 dB above the measured sensitivity level, and then the adjacent channel signal is increased until the FER rises to the same threshold used for sensitivity measurements (i.e., 8% for 802.11b and

10% for 802.11a/g). The difference, expressed in dB, between the two signal levels provides the ACR ratio. Typical ACR ratios range from 0 to 30 dB, and are determined mainly by the quality of the bandpass filters in the receiver downconverter as well as the basic properties of the demodulator itself. The ACR is required by the 802.11 standard to be better than 35 dB for 802.11b, and for 802.11a/g to range between 16 dB at 6 Mb/s and as little as −1 dB for 54 Mb/s. (Note that with OFDM modulation, the signal skirts are so wide that two adjacent channels carrying signals of the same level will cause some mutual interference.)

ACR measurements require high-quality VSGs, that adhere closely to the transmit spectral masks set for the modulation being used. If the VSG providing the adjacent channel signal is of poor quality and does not conform to the spectral mask, the results will be substantially inaccurate and will not correspond to the actual receiver performance. Also, if the VSG signal is much better than the spectral mask, then the ACR ratio measured can be rather optimistic (i.e., not achievable in real practice), because the amount of energy received from the adjacent channel can be lower than anticipated. Generally, ACR ratio measurements should be carried out using high-quality equipment (rather than random off-the-shelf WLAN devices) because of the need to ensure close compliance with the spectral mask.

A related measurement is non-adjacent channel rejection. This is performed in the same way as ACR ratio measurements, except that instead of the interfering VSG being tuned to the immediately adjacent channel, it is tuned to a channel that is somewhat distant from the desired channel. For example, if the desired channel is set at channel 6 in the 2.4 GHz band (i.e., 2437 MHz center frequency), then the interfering VSG would be tuned to either channel 4 (2427 MHz center frequency) or channel 8 (2447 MHz center frequency). Apart from the setting of channel center frequency, the rest of the measurement procedure is identical. Typical non-adjacent channel rejection ratios are 10 to 20 dB better than the ACR ratios, as much less interfering energy makes its way into the receiver passband.

4.3.2 CCA Assessment

IEEE 802.11 requires that the receiver detect the presence of an existing signal within the channel within a specified time after the signal begins (<4 µs for 802.11a and 802.11g in short-slot mode, <25 µs for standard 802.11b, and <15 µs for standard 802.11g). This is referred to as the CCA detect time. Measurement of CCA detect time is important because failure to meet the IEEE 802.11 specification for this parameter can lead to a higher rate of collisions and corrupted frames, due to failure to defer properly to other stations.

CCA detect time is generally measured by configuring a VSG to produce a repeated stream of frames at a specified distance apart (much greater than the CCA duration) and then looking at a CCA detection signal (usually the carrier sense output) with an oscilloscope. The oscilloscope is triggered by the VSA, and the carrier sense signal from the RF/IF or baseband is connected to the vertical input of the oscilloscope. The delay between the trigger point and the carrier sense is the CCA detect time; the measurement cursors of the oscilloscope can be used to find this value. In addition, the VSG should be adjusted to various output levels to measure the CCA detect time as a function of transmit power, which is also a useful metric. In fact, 802.11 specifies a fairly low input level, between -76 and $-80\,$dBm, for CCA sensitivity. CCA should also be measured over all channels and PHY bit rates.

4.3.3 RSSI Accuracy

WLAN receivers measure the signal strength of the incoming received frames and output it to the MAC and upper-layer software as the received signal strength indication, or RSSI. The RSSI measurement is a significant function because it is used for many different purposes (selecting an AP to associate with, adapting the transmit rate up or down, determining when to roam from one AP to another, etc.). Therefore, it is necessary to verify that the RSSI reported to the rest of the system by the receiver RF datapath is as close as possible to the actual strength of the input signal.

The RSSI measurement is generally made by connecting a calibrated VSG to the receiver datapath and then transmitting frames from the VSG to the receiver at a fixed and known signal level. The RSSI found by the DUT is most usually read from the internal RSSI registers within the chipset. The RSSI must be measured only after calibrating and aligning the radio, and ensuring that the AGC and LNA switching is working correctly, as these all affect the measurement that the DUT receiver performs. The measurement is usually made over the entire RSSI range, and also over all channels, to ensure that the RSSI function is linear throughout the operating area.

4.3.4 Total Isotropic Sensitivity

Total isotropic sensitivity (TIS) is the logical inverse of TRP (described above under transmitter testing), and is also a basic system measurement. From a physical point of view, TIS is the sensitivity of the DUT receiver as measured with a perfectly isotropic incoming signal. This effectively integrates and averages out the effects of the DUT antenna pattern, which can otherwise produce widely varying sensitivity figures. Thus, as in the case of TRP, TIS can provide a single figure of merit that is useful for comparing the performance of two different devices with widely varying antenna radiation patterns.

In reality it is nearly impossible to produce a perfectly isotropic signal for use in a TIS measurement; some compromises need to be made. An approximation may be possible with a reverberation chamber, but this has other issues, such as a large amount of delay spread, that makes it difficult to use with WLAN signals. Instead, the customary method of measuring TIS is similar to that for TRP: the DUT is rotated in three dimensions using a 2-axis positioner, the

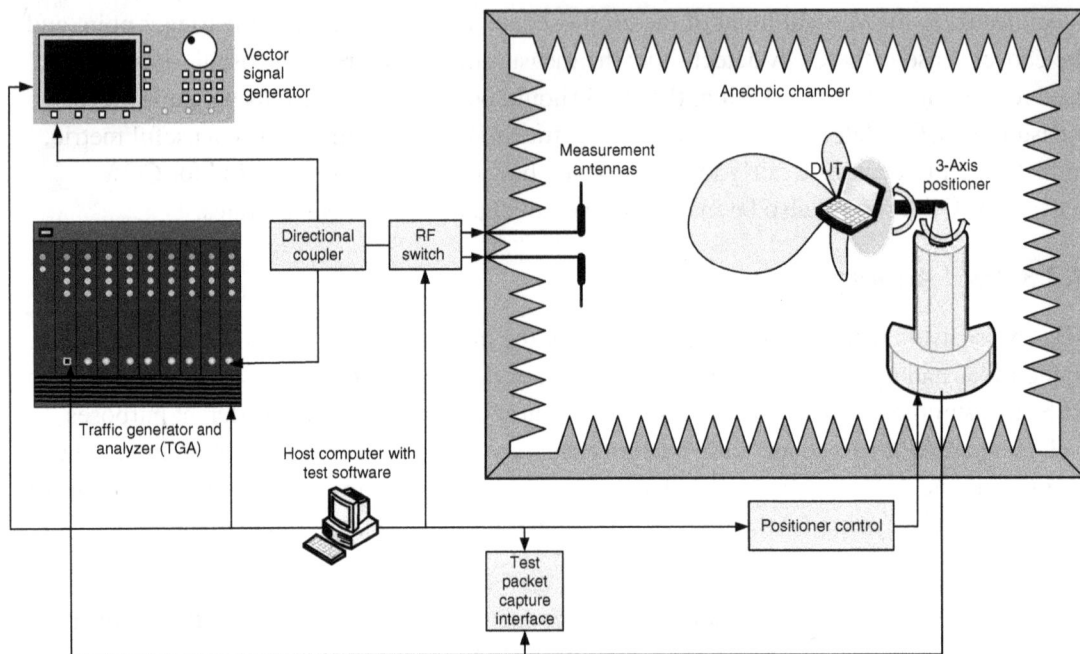

Figure 4.7: TIS Measurement

sensitivity measured for each rotation angle, and the measurements then integrated to obtain the TIS. The figure below shows TIS measurement setup.

The TIS is calculated by inverting the measured sensitivity for each solid angle, integrating over the surface of a sphere, then inverting the result, according to the equation for TRP given above, but substituting (1/TIS) for the parameter T instead.

TIS measurements are usually made at intervals of between 5° and 15°, in order to obtain an accurate figure when considering the radiation patterns observed with WLAN antennas, which commonly have deep nulls and many lobes when installed in the equipment. As with TRP measurements, due to the large number of data points to be obtained and the complexity of the equipment configuration to measure each data point, TIS measurements are automated and run as programs or scripts on a control computer.

4.4 Electromagnetic Compatibility Testing

Besides measuring RF capabilities to ensure that the performance meets datasheet specifications, it is also necessary to ensure that the system meets the applicable regulatory requirements for electromagnetic compatibility (EMC), and the emissions limits for the operating frequency bands.

4.4.1 Regulatory Requirements

WLAN devices must be tested to ensure that they adhere to regulatory limits in the countries in which they are to be marketed. For WLAN devices, this means testing for electromagnetic interference (EMI) limits, as well as testing to ensure compliance to the specific requirements for WLAN devices in the applicable 2.4 and 5 GHz bands. In the US, this means verifying that the device meets Federal Communications Commission (FCC) Class B EMI limits for consumer devices, as well as FCC Part 15 emissions limits for unlicensed intentional radiators.

Note that different regulatory areas specify different limits and requirements. For example, European countries fall under the ETSI ETS 300 standard, while in Japan these limits are defined by TELEC. We will focus here on the US limits, as they are generally representative of typical requirements and specifications. In this case, the limits are set by Part 15 of the FCC Rules.

For WLAN devices, the actual transmitted power from the device is limited by rules given in FCC Part 15.247(b), which limits the transmitter peak output power to no more than 1 W in the 2.4 and 5.8 GHz bands. The 5.15 and 5.25 GHz bands have lower limitations on transmitter power (50 and 250 mW, respectively).

For WLAN devices operating under Part 15 rules, the FCC now enables 'self-certification' via a Declaration of Conformity. This means that the vendor of the Part 15 device must test their device in an accredited emissions testing lab and submit relevant documentation to the FCC, but need not provide the device to the FCC or its associated Telecommunications Certification Bodies (TCBs) in order to obtain FCC approval.

4.4.2 Unwanted Emissions

Most electronic equipment, especially digital devices (including WLAN systems), generate RF emissions over a wide range of frequencies. These emissions are generated by the internal signals of the device; for example, a digital signal switching at 40 MHz generates RF signals at harmonics of 40 MHz. As digital equipment contains a wide variety of signals with many different switching rates, the result is wideband RF emissions. Special design provisions such as shielding, bypassing, filtering, and so on are made in order to hold these emissions under the maximum limits prescribed by the FCC. The measurement and verification of compliance to these limits is known as EMC or emissions testing.

Emissions testing is normally done with over-the-air measurements: first, the FCC regulations typically specify the field strength at a distance of 3 m from the DUT; and secondly, the unwanted emissions can take place from parts of the DUT other than the actual RF components, so cabling to the DUT is not possible. Both conducted and radiated emissions are measured during the tests.

For radiated emissions, an FCC-specified emissions mask is used, as given in FCC Part 15.247(c). The test distance is 3 m, and the mask range covers the frequencies from 1.7 MHz to 1 GHz. In this range, the mask field strength limits are as follows:

Frequency range	Signal level
1.705–30 MHz	30 μV/m*
30–88 MHz	100 μV/m
88–216 MHz	150 μV/m
216–960 MHz	200 μV/m
960–1000 MHz	500 μV/m

*For reference, a 100 μV/m field strength at 3 m corresponds to an isotropic radiated power of about -55 dBm.

The FCC requires that ANSI C63.4 ('Methods of Measurement of Radio-Noise Emissions from Low-Voltage Electrical and Electronic Equipment in the Range of 9 kHz to 40 GHz') be used as the test methodology above 30 MHz.

For conducted emissions, CISPR 22 is used as the limit specification. (CISPR stands for 'Special International Committee on Radio Interference,' the acronym derives from the French name, and is an International Electrotechnical Commission special committee formed to standardize limits and measurement methods for electromagnetic compatibility.) CISPR 22 specifies a maximum conducted signal of 631 μV for frequencies between 0.15 and 5 MHz, and 1000 μV between 5 and 30 MHz, measured using a 'quasi-peak' method. Further, FCC Part 15.207(a) specifies that the average signal conducted back on to the AC power line must not exceed 250 μV in the 450 kHz to 30 MHz band.

A sensitive spectrum analyzer and a measurement antenna (cabled to the analyzer for flexibility) are generally used for radiated emissions testing. A field strength meter may be present as well, but the spectrum analyzer is required for determining mask compliance. The DUT and measurement antenna are placed within a well-isolated anechoic chamber to remove any external interference, or else the testing is conducted on an open-air antenna range. A complex three-dimensional positioner is not necessary; instead, a frequency sweep is taken at one or two different measurement antenna locations and orientations relative to the DUT to ensure that the worst-case emissions are being measured. Different configurations of the DUT are measured: cables attached, cables off (except the power cable), etc. The configuration that must meet the emissions mask is the manufacturer's recommended usage configuration (i.e., the configuration that customers of the product are expected to use).

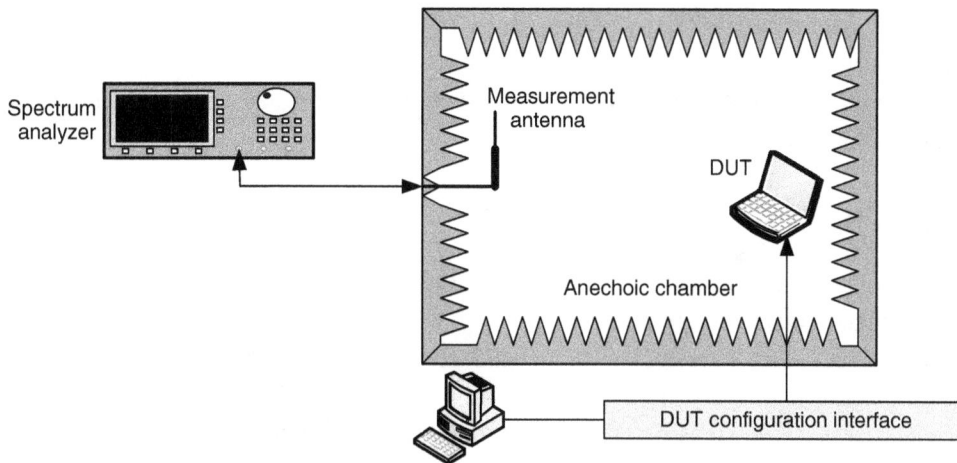

Figure 4.8: Emissions Test Setup

Conducted emissions testing is done with a spectrum analyzer and a transducer, essentially a high-bandwidth current transformer, that picks up spurious emissions being conducted down cables attached to the DUT. The power cables, digital signal leads, and even RF cables are all expected to be tested for conducted emissions. Note that CISPR 22 testing requires a special spectrum analyzer with quasi-peak detectors.

4.4.3 Spectral Mask Compliance

The modulation formats (DSSS, CCK, OFDM) used by IEEE 802.11 generate very wideband signals between 16 and 20 MHz in terms of -3 dB bandwidth. To ensure that they minimize or limit cross-channel and out-of-band interference, IEEE 802.11 transmitters are required to adhere to the appropriate emitted power spectral density vs. frequency characteristics, as defined by a spectral mask (see Chapter 1). Spectral mask compliance is measured with a spectrum analyzer; many lab-quality spectrum analyzers support software packages that superimpose the 802.11 spectral mask corresponding to the frequency band and modulation type in use on the displayed signal spectrum, making it simple to determine whether the device meets spectral mask limits. Failure to meet spectral mask limits usually indicates distortion in the RF chain or malfunctioning filters. For example, a distorting PA leads to 'spectral regrowth,' which is basically the generation of unwanted sidebands due to nonlinearities in the PA.

Measurement is straightforward if a spectrum analyzer with the appropriate mask software is available: simply select the mask option, configure trigger parameters, cause the system to transmit frames, and verify that the displayed spectrum falls entirely within the mask. The IEEE 802.11 standard specifies that the spectrum analyzer RBW when making spectral mask measurements should be 100 kHz, and the video bandwidth should be set to 30 kHz (except for 802.11b, where the video bandwidth is 100 kHz). No specification is made as to the type of frames used, but ideally they should be data frames of maximum size and containing random data, to provide a reasonable approximation of a worst-case situation.

4.4.4 Radar Detection

In certain regulatory areas (most notably Europe), the 5 GHz band is assigned on a primary basis to aerospace radars. To avoid interference by 802.11a WLAN devices (which are unlicensed and secondary users of the same band), ETSI mandates that such devices should attempt to detect these radars, and, if detected, shift to a channel that is not occupied by the radar. This process is known as radar detection and is an important compliance parameter that must be designed in and verified before the equipment can be sold into such regulatory areas.

The specific method of radar detection is not mandated by the 802.11 standard; it explicitly leaves the actual implementation up to chipset and system vendors, only stipulating that radar detection must be performed. However, the usual process is to attempt to detect, in the baseband of the receive RF datapath, energy above a certain threshold (-51 dBm, per ETSI rules), and then to verify that the detected energy does not resemble a valid 802.11a preamble or frame. The detection is performed during 'silence periods'; these may be forcibly inserted as per the 802.11 radar detection protocol, or may be the gaps between frames (e.g., the SIFS or DIFS periods), or both. If energy not corresponding to a portion of valid 802.11a frame has been detected over a certain number of averaging intervals, the baseband signals to the MAC that radar has been detected and the system must move to a different channel.

Radar detection testing is ideally performed by setting up a signal source to mimic the spectral characteristics of the aerospace radars, possibly even using the actual radars themselves, but this is understandably rather difficult! Instead, a simple expedient is to simulate the radar signal with a pulsed RF signal generator. This consists of a standard signal generator gated by an external pulse generator; virtually all laboratory signal generators support this function. The signal generator generates a continuous (CW) RF signal in the 5.15–5.85 GHz range, and the pulse generator imposes an on/off keying or modulation on this signal to produce a series of short pulses. The pulse widths should be limited to 0.1 µs, and the pulse repetition frequency to a few kilohertz. The peak output power of the resulting signal should be adjusted to the radar detection threshold of -51 dBm and then applied to the DUT on the same channel to which it is tuned. If the DUT baseband indicates that a radar has been detected, then the measurement is considered to have succeeded. For a more complex measurement, this process should be repeated, but with data frames being injected into the DUT at the same time as the pulsed signal generator output is applied (via a power combiner).

4.5 System Performance Tests

Some aspects of the PHY layer, such as rate selection to minimize FER, are implemented in conjunction with the MAC functions and even the device driver or operating firmware. They can thus can only be verified using system-level tests; that is, tests on the complete system, with all components integrated and running the expected operating firmware. This subsection therefore treats typical system-level tests.

4.5.1 Rate vs. Range or Path Loss

It has been observed by every 802.11 user that the achievable effective transfer rate of an 802.11 link within a given environment depends quite significantly on the distance between the AP and its associated client. As the distance increases, the signal strength at the receivers on each end of the link drops off; this is due both to the reduction in field strength as the distance from the transmit antenna increases, the increased number of attenuating elements (walls, furniture, etc.), and increased multipath between the two ends of the link. The consequence is that the signal-to-noise (SNR) ratio falls, and bit errors rise sharply for a given modulation type. The reduced SNR causes the system to drop its PHY bit rates (see below) to maintain efficient data transfer, and also causes an increase in retransmissions. The user-visible effect is thus a drop-off in application layer network performance, caused by the reduction in overall data transfer rate of the 802.11 link.

As this drop-off of transfer rate determines the usable coverage of the 802.11 AP, it is of significant interest as a performance metric. Unfortunately it is not very easy to measure, because it is highly dependent on the environment. For example, a building with a higher density of absorbing materials (e.g., one with more walls per unit area) will cause a faster drop-off than a relatively open building. Thus a measurement of rate as a function of distance between client and AP in a real building is not valid for anything other than that particular building, and usually is not even valid within that particular building for anything except the points selected for the measurement.

Instead, the common practice is to measure the transfer rate profile of the AP in an idealized scenario such as an open-air or a conducted environment, and then later map this profile to a rate physically achievable in a given building environment by factoring in the actual absorbers within the building. (The propagation modeling tools described in a subsequent chapter can be used in this regard.)

Two different setups are applicable to the measurement of this metric: a well-characterized open-air environment (e.g., an outdoor antenna range) or a fully conducted environment. The open-air test setup measures the transfer rate in terms of range directly; that is, it produces the variation of data transfer rate with the distance between the measurement antenna and the DUT. The conducted environment measures the rate vs. range function indirectly, by determining the transfer rate as a function of the path loss inserted into the RF path between the test equipment and the DUT. In the latter case, the path loss can then be used to estimate the range in a given environment, provided that the properties of the environment (attenuation, multipath, etc.) are known. A propagation modeling software package, for example, can be used to determine the path loss between any two points in a building; the rate vs. path loss function then immediately yields the expected 802.11 transfer rate between those points. The two different setups are illustrated in the figure below.

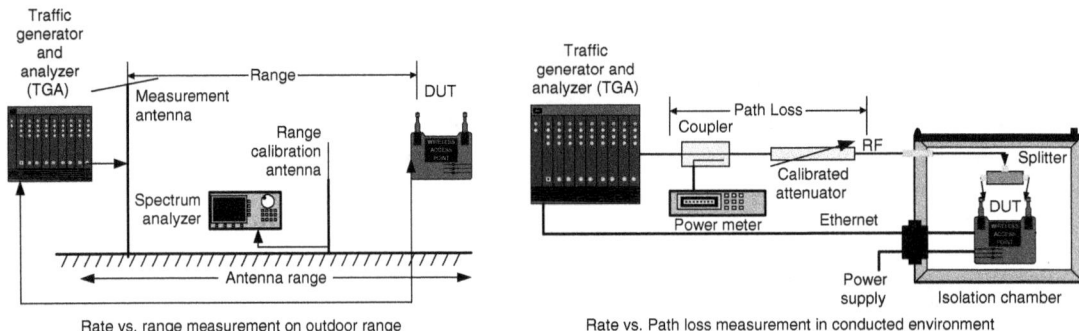

Rate vs. range measurement on outdoor range

Rate vs. Path loss measurement in conducted environment

Figure 4.9: Rate vs. Range Testing

In the open-air version of the test, a traffic generator of some type is used to exchange a stream of data packets with the DUT. The distance between the DUT and the traffic generator is progressively increased and the goodput (i.e., number of 802.11 data frames successfully delivered per second) is recorded for each value of distance. This produces the rate vs. range function for a free-space environment (assuming that ground reflection can be neglected). The Friis transmission equation, which is as follows:

$$P_r = P_t \times G_t\, G_r\, \lambda^2/(4\pi r)^2$$

where:
P_r = received power
P_t = transmitted power
G_t = gain of transmit antenna
G_r = gain of receive antenna
r = distance between antennas (range) and λ is the wavelength

can then be used to convert the free-space range into a path loss. The path loss as a function of range is simply:

$$Path\ loss\ (dB) = 10\ log_{10}[G_t\, G_r\, \lambda^2/(4\pi r)^2]$$

As noted, once the rate vs. path loss is known, propagation modeling software can be used to estimate the rate in the actual environment.

The conducted form of the test interposes a path loss directly, as a variable attenuator. In this case, the attenuator is merely varied in steps and the transfer rate between the AP and the traffic generator is noted at each step. The fixed path loss of the rest of the components (coupler, splitter, cables, etc.) must be factored in as well. This directly produces the rate vs. path loss function, which is used as described above.

Note that the rate vs. path loss (or range) function can also be used in a comparative sense, to determine which of the two different APs would produce a larger coverage area for the same power setting. Obviously, the AP that maintains a higher rate for a given path loss should produce a larger coverage area.

One problem with both of the above approaches is that the measurement is not very repeatable, as it depends highly on the manufacturing tolerances of the radio used in the client (or traffic generator acting as a client). Testing the same AP with two different instances of the same make and model of client, from different production batches, produces quite different results. Recently, however, VeriWave Inc. has introduced a novel variation of the rate vs. path loss test to overcome this issue when testing APs. Basically, this observes that, from the AP's point of view, an increased path loss (or range) manifests itself as two concurrent phenomena:

1. An increased number of retries in frames transmitted by the AP, due to bit errors being caused by a lower SNR at the client.

2. A reduced received signal strength for frames received by the AP from the client (which in turn causes a lower SNR and higher BER at the AP).

The first effect is manifested by a failure of the client to return 802.11 acknowledgement frames in response to valid data frames transmitted by the AP. As acknowledgement frames (or lack thereof) form the only indication to an AP of the SNR (and thus the BER) situation at the client, it is possible to simulate an increasing SNR at the client by deliberately withholding acknowledgements to a certain fraction of data frames. The AP interprets the lack of an acknowledgement as a bit error at the client, retransmits the frame, and also starts reducing its PHY rate. This is exactly equivalent to the situation where an increased path loss is introduced between the AP and the client. The rate at which acknowledgements are withheld is exactly equal to the perceived BER at the client (i.e., the traffic generator acting as a client).

The second effect is also equivalent to an increasing path loss, except that rather than introducing an attenuator or physically moving devices apart, the client (or rather the traffic generator acting as a client) steps down its transmit power. Lowering the transmit power of the client is exactly analogous, from the AP's point of view, to increasing the attenuation or increasing the range.

The two effects can be linked, so that they can act concurrently, by noting that the relationship of BER to SNR for a given modulation type is well known. The procedure for conducting the test is then as follows:

1. Increase the (virtual) path loss.

2. Calculate the SNR expected to be present at the client for this new value of path loss.

3. Calculate the BER resulting from this SNR for the given modulation type.

4. Withhold acknowledgements at a rate equal to the BER; if the BER is calculated to be 10%, for example, withhold acknowledgements for 10% of the data packets received from the AP.

5. Measure the goodput, and then reduce the traffic generator transmit power by an amount equal to the increase in path loss.

6. Repeat the above steps until the entire path loss sweep has been completed.

When the two effects are introduced simultaneously, in the above manner, it is found that the net effect on data transfer rate (as well as the rate adaptation characteristics of the AP) are identical to the actual variations observed when the range between the AP and the traffic generator increases. The great advantage of conducting the test in this way is that the manufacturing tolerances of the tester or client radio do not affect the results; the measurements reflect only the properties of the AP. This greatly improves the repeatability and reproducibility of the test.

4.5.2 Receive Diversity

Many WLAN devices, even clients built into laptops and handhelds, support diversity reception. Diversity will be described in more detail in Chapter 9, but briefly speaking it is a technique for improving signal reception in a fading environment by employing multiple antennas and selecting the best signal received from these antennas. The selection is done during the preamble of each received frame; the WLAN receiver simultaneously measures the RSSI at each antenna, and uses an RF switch to select the antenna providing the maximum RSSI. Many different forms of diversity are known: space diversity, pattern diversity, polarization diversity, etc. However, the most common configuration for a WLAN device is space diversity, using two identically polarized antennas spaced a small distance apart (typically ½λ).

Note that diversity transmission is also possible and often implemented; the best receive antenna is also the best transmit antenna for communicating with that particular remote station.

A vendor-specific algorithm is used to measure the RSSI, determine the best antenna to use and switch to the antenna without losing so much of the preamble that the frame cannot be decoded. The performance of this algorithm materially impacts the overall throughput and stability of the WLAN DUT, and hence it is important to quantify this.

Diversity is best measured in a conducted environment, where differential signal strengths can be provided to the two DUT antennas in order to simulate the situation in a small-scale fading scenario. This is illustrated in the following figure. Note that it is assumed that the DUT antennas are removable and connectorized, which is true in most cases where diversity is employed.

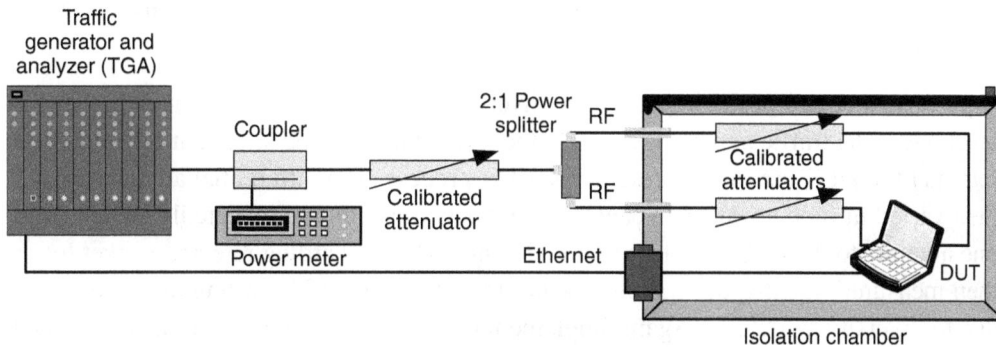

Figure 4.10: Diversity Test Setup

As shown in the figure above, a set of variable attenuators is used: a separate attenuator per antenna input of the DUT, plus a common attenuator in the signal path. The attenuators connected to the antenna inputs of the DUT are used to provide differential signal intensities to the two DUT antenna ports, thereby triggering the diversity algorithm. The common attenuator is used only to establish a baseline signal level that is convenient to perform the test, and to reduce the range required from the differential attenuators. A traffic generator is used to transmit packets to the DUT (and handle responses).

A power meter with an associated directional coupler are employed to measure the absolute power being driven from the traffic generator, so that the actual power levels applied to the DUT can be calculated using the path loss in each leg. All of the attenuators are required to be calibrated, and the losses in the cable, coupler, and power divider must be known; this ensures that the path loss through any leg of the system is known for all settings of the attenuator. (It is advisable to disconnect the cables from the traffic generator and the DUT antenna connectors and use a network analyzer to measure the path loss in each leg with the attenuators set to the minimum value; this provides the most accurate means of calculating the absolute power at the DUT terminals.)

Two kinds of tests are possible with this setup:

1. A functional test of the diversity switching algorithm at the system level. In this case, the traffic generator sends a steady stream of packets, and the differential attenuators

are varied so as to reduce the power applied to one DUT antenna port while increasing the power applied to the other port. The antenna selector switch control, or the antenna selection status as reported to the system software, is monitored. At some differential attenuation, the antenna selection will change; the input power difference required to cause diversity switching can be calculated directly from the difference in attenuator values.

Note that once this level is known, the differential attenuator can be varied in the opposite way, and a second switching point noted. Ideally, the two switching points should be the same, but implementation issues in the diversity switching algorithm may cause one antenna to be 'preferred' over the other.

2. A performance test of the impact of the diversity switching algorithm, again at the system level. Here the traffic generator presents the DUT with a traffic stream at the maximum rate that the DUT can successfully forward or process. The differential attenuators are then varied in such a way as to cause the diversity algorithm to repeatedly switch from one antenna to the other, while traffic is flowing. The forwarding rate and packet loss are then measured. Ideally, there should be no impact on either forwarding rate or packet loss due to diversity switching. Again, implementation issues may cause a reduction in DUT capacity while diversity switching is occurring.

    A variant of this test attempts to compare the maximum forwarding rate supported by each of the two DUT antennas. The procedure is the same, but instead of repeatedly switching the DUT antennas back and forth, a switch is caused to one antenna, the maximum forwarding rate is measured, a switch is caused to the other antenna, and the maximum forwarding rate is measured again. Ideally the two forwarding rates should be the same, but if RF design problems exist in the two antenna paths, there may be differences.

Isolation and cross-coupling are the significant issues to guard against when measuring diversity performance. Typical diversity switching thresholds (i.e., the power difference between two antennas that will cause the diversity switching algorithm to shift from one antenna to the other) are normally on the order of 10–20 dB, which is quite substantial. Some diversity switching algorithms also have absolute power thresholds built in for stability, and so will not consider switching until the RSSI on the lower-signal antenna drops below some predefined level (usually −40 to −60 dBm). This means that the attenuation values may have to be quite large – 50 dB in the common leg is not unusual – in order to cause diversity switching to occur. Attenuator 'blow-by,' that is, the leakage of signal power past the attenuator, causing an upper limit on the maximum attenuation value, is therefore an issue to be addressed. Further, leakage from the cables can cause coupling across the two differential legs, which can limit the maximum differential power that can be applied to the DUT antenna inputs. (The use of high-quality cables and SMA connectors can more or less eliminate this issue.) Also, chamber isolation should be checked carefully, as the signal difference between the traffic source and the DUT antenna terminals can be on the order of 70 dB or more.

4.6 Getting the DUT to Respond

A basic issue with the testing of transmitter and receiver submodules is that, as they are not yet complete systems with the accompanying firmware and hardware that makes them work as stand-alone units, signals must be artificially injected and extracted to get the DUT to 'do something.' Without a comprehensive and straightforward method of causing signals to flow through the DUT, it is not possible to examine its responses and determine if the DUT is working correctly. Several different options exist for carrying this out.

Note that one requirement is common to all the different modes of injecting signals into and extracting signals from the DUT: its internal registers must be configured with the proper data before it will even begin to function. The configuration interface is most commonly constructed with the aid of general purpose I/O pins driven from a PC, as this allows the register contents to be easily changed by modifying the script or program that configures them via the I/O pins.

4.6.1 External Signal Injection Points

One obvious means of driving signals into and out of the DUT is to connect directly to its external connectors and artificially generate the signals necessary. This is shown, for example, in Figure 4.1, where a bit pattern generator (or similar setup using a PC) is used to produce the bit sequence that the baseband device interprets as frame data to be transmitted. Once this is successfully set up, the baseband will drive the RF components, and the response of the entire module can be measured by a VSA, for example.

4.6.1.1 Probing

A significant and complex area that has not been touched upon in the foregoing discussion is the area of probing. In the event that the individual components within a module are to be separately tested or monitored, it is necessary to probe signal pins or traces. Probing a microwave subsystem is a complicated and time-consuming task, for two reasons:

1. Probes introduce discontinuities and mismatches into signal paths, which become quite substantial at microwave frequencies.

2. The modules and interconnects themselves are quite small and difficult to probe, and the probes used are delicate and expensive (and have to be used under a microscope).

For these reasons, RF design engineers spend a great deal of time and money selecting probe points, devising probe structures and acquiring high-frequency probe hardware. However, these issues are mostly attendant upon development testing, when the components of a module need to be tested and verified. Once the module has been fully developed and connectorized, probing is no longer necessary; instead, standard cables with adaptors can be used to connect the test equipment to the RF signal ports. In fact, at that point it may actually become more time consuming to gain access to the digital signal ports (which can involve high-density connectors and software development) than to the RF signal inputs and outputs.

4.6.2 Internal Test Modes

Most WLAN chips (except pure analog devices) include some sort of test mode designed primarily to enable the IC verification engineers to test the chips in the absence of a complete system. These test modes can frequently be used in module and even system test as well, after the chipset has been integrated onto a board. The test modes are usually accessed through special registers on the chip(s), though in some cases special test buses are also made available for such purposes. In most situations the test modes can only be utilized if the normal operational mode is disabled or bypassed.

A number of different test modes can be useful for testing signal flows through modules and verifying transmitter or receiver performance. Unfortunately, the availability and nature of these test modes vary widely between chipset vendors, and in fact even vary between different chipsets from the same manufacturer. The chipset datasheets have to be consulted before the possibility of using a test mode to cause the chipset to transmit or receive data.

Some examples are as follows:

1. The transmit baseband may be capable of entering various test modes, such as a mode for generating pseudo-random bit sequences (PRBS) and injecting them into the transmit chain. These PRBS sequences are usually continuous bitstreams that take the place of transmitted frames. PRBS sequences are particularly useful for measuring output spectrum and average transmitted power, and producing a test signal for evaluating RF/IF components before anything else is available to drive the baseband. (In fact, the original 1 and 2 Mb/s DSSS 802.11 standard specified several mandatory test modes that had to be implemented by PHY vendors; unfortunately these modes are not required for more modern 802.11b/g and 802.11a PHYs.)

    In many cases, even if a PRBS generator is not available, it is possible to force the chipset to transmit a continuous carrier for the purpose of tuning during production calibration – this is also usable for test purposes.

2. Some devices offer the option of placing a single frame into a buffer and then initiating a looped transmit (i.e., transmit the same frame over and over again). This is generally available only when the MAC is part of the chipset. Such a test facility allows a wide variety of tests, including EVM measurements and spectral mask compliance, to be conducted without requiring external sources of transmit frames. If this is not available (or accessible), it is possible to perform a similar function with a somewhat higher degree of effort by utilizing the on-chip DMA capability and on-chip memory buffers on most MAC devices to store a user-defined frame and repetitively cycle the frame through the transmit

path. In addition, it is possible to capture one or more incoming frames into the on-chip RAM for subsequent inspection by software or scripts, which is a useful function for receive datapath testing.

3. Most devices contain test buses that enable IC design and verification engineers to get at the internals of the chips and do debug and device test functions. These test buses may be made accessible by the manufacturer, in which case they offer a considerable range of functions for both transmit and receive testing.

4. If all else fails, most chipsets include an embedded CPU of some kind, and software development kits (SDKs) for the embedded CPU are available from the manufacturer. In this case, a test program can be developed and uploaded into the embedded CPU to transmit and receive test frames.

4.6.3 Host Test Software

In the case of a client NIC (Network Interface Card) module, the simplest way of exercising the RF and baseband functions is to insert the module into a desktop host computer (e.g., using a PCI-bus adapter for a PCCard or PCMCIA module) and create software on the host that drives the module. In some cases it may be necessary to modify portions of the standard NIC driver to place the devices into the desired operating modes. A similar approach can be used for APs, but in this case a straightforward setup utilizes the embedded CPU used for the AP rather than a separate host computer.

4.6.4 Packet Traffic

If the MAC is even partially functional, then by far the simplest way of getting frames into and out of the DUT is to use an external traffic generator and transmit frames to the DUT. For example, most AP MACs will issue beacons and respond to probe requests, provided the chipset registers have been set up correctly. In this case, using an external traffic generator to send a continuous stream of probe requests will cause the chipset to output a corresponding stream of probe responses. Various similar tricks can be used, depending on the level of functioning of the MAC device. This approach is most commonly used for APs; getting client devices to function in this manner is much more difficult. Ping (ICMP echo request) frames are frequently used when dealing with clients, if the driver and IP stack are functional.

Care should be taken to ensure that the frames transmitted by the DUT are of the desired PHY bit rate. For example, beacons and probe responses are management frames and typically transmitted at low rates (1 Mb/s for 802.11b/g and 6 Mb/s for 802.11a), making them unsuitable for measurements such as EVM. In this case the only recourse is to force the DUT to send data frames with whatever means may be available, even down to programming frames bit-by-bit into a digital pattern generator.

If frame traffic is sent to the DUT by an external traffic generator, then care must be taken to avoid inadvertently including the signals from the traffic generator into the measurement. A simple method of achieving this is to ensure that the signals from the traffic generator are significantly attenuated relative to the signals from the DUT at the input port of the measurement instrument (e.g., at least 30 dB below that of the DUT), and then setting trigger levels to prevent triggering on the traffic generator frames.

Protocol Testing

Metrics and measurements pertinent to the wireless LAN (WLAN) Medium Access Control (MAC), as well as the Transmission Control Protocol (TCP)/Internet Protocol (IP) stack, are covered here. Protocol testing covers a wide swath of measurements: performance, conformance, functional and interoperability, as well as comparative benchmark testing. Most of the focus in this chapter is on WLAN MAC-level testing, including security.

5.1 An Introduction to Protocol Testing

While technically the (RF) physical or PHY layer is considered a part of the WLAN protocol, in this context protocol testing refers to tests performed on device functions involving frames. Typically, when protocol tests are performed the PHY layer is assumed to be working and standards-compliant, and therefore no attempt is made to include it in the measurements. Thus protocol tests are aimed at the link layer (and above) of the ISO protocol stack, as indicated in the figure below.

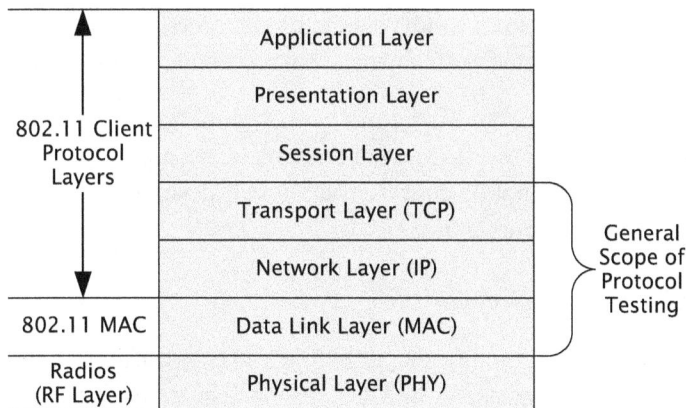

Figure 5.1: ISO Protocol Stack

5.1.1 Functional vs. Performance Testing

It is not unusual to find functional tests being confused for performance measurements when performing protocol testing. However, the difference is important to understand.

A functional test is concerned with verifying that a device or system does something. For instance, a functional test might check to see that a WLAN device transmits frames that have properly formatted MAC headers and are without Frame Check Sequence (FCS) errors. (Conformance testing is a close cousin to functional testing, in that the test results are also usually binary – yes/no.) Even if a functional test produces numbers as the output result, a "correct" number is generally known in advance, and the test results can be compared to that number. Thus for example if a WLAN access point (AP) is expected to accept packets ranging from 28 to 2346 bytes, then a functional test would inject packets of different sizes and ensure that it could in fact support this range.

Functional tests are also quite narrowly focused and specific. Frequently they are aimed at verifying some particular feature of a device, or whether the device supports a specific requirement of the IEEE 802.11 protocol. Marketing people rarely have much interest in the results of functional tests; they are of most value to development and Quality Assurance (QA) engineers.

A performance test is much more broad-based and generic, and is aimed at characterizing a device or system according to some well-defined metric. Indeed, a single performance metric may be applicable to a wide variety of devices (not merely WLAN devices); for example, the throughput performance metric can be applied to Ethernet switches and DSL modems just as easily as to WLAN APs. There are no yes/no answers, and the results of performance tests are often only interesting when compared to the results of similar performance tests performed on other (often competing) devices. The marketing department therefore takes great interest in the output of a well-formulated performance test. QA personnel may also use performance tests to verify that the device is up to snuff, but the engineering department usually does not – they can often predict the expected performance characteristics of the device without even bothering to run the test.

Thus, in a nutshell, a functional test is directed at correctness; a performance test is directed at goodness. These two should not be confused. If a test is designed to prove something, then that test is almost always a functional test.

5.1.2 Test Stimuli And Measurements

WLAN protocol tests, almost without exception, inject packet traffic into the device under test (DUT) as test stimuli, and measure the DUT response in terms of the number or rate of specific types of packets that it generates in turn. As tests are being performed on the MAC layer, the RF characteristics of the packet traffic are rarely of much interest, beyond ensuring that these characteristics do not skew the test results by causing unexpected issues at the PHY layer. It is usually preferable to set signal levels and signal quality such that both the DUT and the test equipment have no difficulty in receiving normal traffic. Thus, for example, the signal levels of the packets injected into the DUT by a tester should be placed within the dynamic range of the DUT receiver to strike the best compromise between signal-to-noise ratio (SNR) and intermodulation distortion (IP2 and IP3), and hence minimize the number of random bit errors produced within the DUT.

An exception to this occurs when some PHY layer effect causes MAC or upper layer protocol behavior to change. For example, a gradual decrease in signal strength can indicate to an AP that a wireless LAN client is moving away from it and may decide to roam to another AP at some point. Such special situations require that the tester combine some PHY layer control and functionality into protocol testing. Note that it is very rare to find wired LAN tests requiring similar PHY layer control as part of protocol testing; this is a problem specific to WLANs.

5.2 Conformance and Functional Testing

As has been explained previously, a conformance or functional test is aimed at verifying that a device conforms to some requirement or implements some feature.

The term "conformance test" is usually employed in conjunction with some official industry standard or de-facto market requirement. For example, tests designed to prove that a wireless LAN device meets the requirements of the IEEE 802.11 standard are conformance tests. Obviously there are no "typical" conformance metrics; a standard as complex as IEEE 802.11 has a tremendous number of aspects to which conformance must be tested. An example of a conformance metric is determining whether a device retransmits correctly if the receiver fails to acknowledge its packets within a certain time.

On the other hand, functional tests are manufacturer-specific; they have little to do with industry standards, and much more with what is on the datasheets of the device or system. A test intended to show that a device supports some feature claimed on its datasheet is a functional test. An example could be a test to verify that an 802.11 chipset can handle both AES-CCMP and TKIP encryption modes.

Test equipment for conformance testing can also be used for functional testing, and vice versa. Two general approaches exist in the industry for such test equipment: stack-based and scripted. A stack-based tester implements a full protocol stack (e.g., the IEEE 802.11 PHY and MAC layers) to allow it to exchange test traffic with the DUT. However, the protocol stack is specially instrumented to enable the user to carry out point measurements, mostly of functionality. A scripted tester, on the other hand, does not have a complete protocol stack, but instead supports a powerful scripting language that allows the user to programmatically generate sequences of test packets and analyze DUT responses. Scripted testers are far more flexible than stack-based testers (since virtually any aspect of the protocol can be controlled by the user), but are correspondingly much harder to use. They are typically utilized for conformance testing, where it is frequently necessary to generate abnormal or illegal

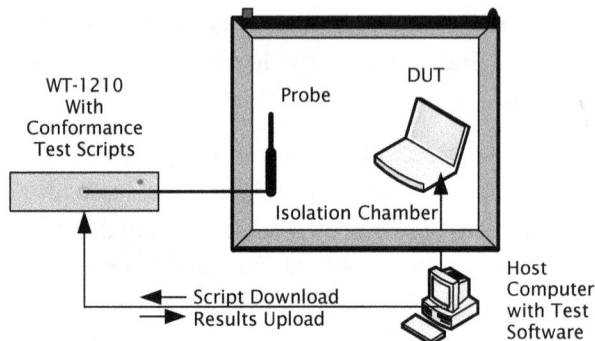

Figure 5.2: Conformance Test Setup with WT-1210

sequences of packets to check that the DUT responds appropriately (and according to the protocol). An example of a scripted WLAN tester is the VeriWave WT-1210 system.

5.2.1 The 802.11 PICS and Conformance Tests

All current IEEE 802 standards make life simpler for a conformance tester by defining a section called the Protocol Implementation Conformance Statement or PICS. As described in Chapter 1, the PICS is a large and detailed table that identifies all of the elements of a standard that an implementer must adhere to in order to claim conformance to that standard. For example, the WLAN PICS is contained within Annex A of the IEEE 802.11 standard, and covers 27 pages. The PICS tables are required to be filled out and supplied on demand by a vendor of a device or system that claims conformance to the relevant portions of the standard.

A PICS is very useful for structuring a conformance test, or building a conformance tester. Searching through a large standard for functions that are mandatory, and hence must be tested, is both laborious and error-prone. Further, determining exactly what is mandatory for a specific implementation is sometimes subjective and open to interpretation. The PICS solves all of these issues. An implementer of a set of conformance tests for a WLAN device only needs to ensure that he or she has created a test for each entry in the PICS tables in the IEEE 802.11 standard. Moreover, the user of a conformance test suite has an immediate cross-reference to the relevant portions of the IEEE 802.11 standard that underlie each test, as a benefit of linking each test to the PICS. For this reason, implementers of IEEE 802.11 conformance tests (or conformance tests for any other 802 networking standard, for that matter) reference their test procedures to the PICS tables.

A PICS normally covers all of the elements of a standard that an implementer must include, or adhere to, in order to claim conformance. For instance, the 802.11 PICS specifies that all 802.11 stations or APs must support Open System authentication; in fact, this is essential for interoperability. Thus a conformance tester can design a test to verify that a device actually supports Open System authentication; if the device fails the test, it can be marked as non-compliant to the 802.11 standard without further ado.

A PICS does not cover implementation-specific details, or informative portions of the standard that are provided as guidelines and not specific requirements. Further, a PICS does not try to mandate performance. Attributes such as the power consumed by an 802.11 device, the throughput available from it, or the latency through it, are not part of the PICS and should not be part of an 802.11 conformance test.

PICS tables for IEEE 802 standards are fairly complex and employ a curious language to express various conditions and predicates. The language is often described in the PICS clauses or annexes themselves, such as in Annex A of IEEE 802.11. The PICS tables need to cover situations where a portion of the standard may be optional (hence the conditional notation) but if that portion is actually implemented, then certain other portions become mandatory (hence the predicate notation). Further, there may be exceptions or exclusions that also result from optional portions of the specification.

5.2.2 Test Methodology

Functional as well as conformance test procedures are meticulous, exacting, and detail-oriented. (That is, if they are properly done!)

Conformance tests are almost exclusively performed in a highly controlled and isolated environment, as external interference at the wrong moment can invalidate an entire test. Further, the DUT is a single device or chip, rather than being an entire system, as the IEEE 802.11 standard covers a single WLAN station.

Functional tests are less intolerant of interference effects, but nevertheless are better performed in a controlled environment from the point of view of repeatability and accuracy. (This is even more desirable in view of the fact that groups of functional tests are often gathered into regression test suites and automated, so that they can be run whenever a hardware or firmware change is made.) Either a single device or an entire system can be tested as a unit.

The general test setup for both conformance and functional testing is usually very simple, often consisting simply of a tester and a DUT. Figure 5.2 shows an example. A shielded enclosure is used to isolate the DUT, and cables connect the DUT to the tester.

One of the unexpected issues that crops up when testing client DUTs is persuading them to generate or accept packet traffic. In most cases, client DUTs are not set up to source or sink traffic unless a user – typically, a human – is doing something with them, such as transferring a file or downloading e-mail. However, involving a human as part of the test equipment is both tedious and also not conducive to repeatable or predictable test results. (What happens if the user presses the wrong key at the wrong time?) Thus special methods have to be used to get such a DUT to accept or transmit traffic so that measurements can be made. A variety of methods are used in practice:

- If the DUT supports a TCP/IP stack, the "ping" (RFC 792 ICMP echo request/ response) protocol can be used to get it to exchange traffic at relatively low rates. Simply send the DUT a ping request, and it should return a ping response. Ping is particularly useful in that the echo response is expected to contain the same ICMP payload as the echo request; hence the DUT can be forced to transmit packets of various sizes by merely sending it to the corresponding size of echo request.

- A special software application can be created and hosted on the DUT, the sole purpose of the application being to source or sink network traffic from the wireless network adapter. This is obviously far more flexible than the ping method, but also requires meddling with the DUT setup and being able to run special-purpose applications on it.

- In some situations, it may be possible to use control and management frames that are part of the 802.11 protocol, and to which the DUT is required to respond. For example, an RTS packet received by the DUT requires that a CTS packet be transmitted in response; a data packet requires an acknowledgement packet (ACK) to be sent in response.

- Spoofing is also possible, though somewhat complicated. If the DUT is associated with a normal AP (in front of a standard network) and then a long file transfer is started, the tester can inject test frames in the middle of normal data transfers to conduct functional or conformance tests. This requires a considerable amount of intelligence on the part of the tester (which now has to follow complete protocols).

APs are much simpler to test in this regard, as they have two ports, and are required to act as bridges. Merely injecting test traffic into one of the ports should cause it to automatically be retransmitted out the other port, and vice versa. Testing the wireless port of an AP thus entails the generation of wireless and Ethernet traffic within the tester.

5.2.3 Conformance Tests

IEEE 802.11 protocol conformance tests are performed on the different aspects of the WLAN-MAC and PHY protocol specifications. An exhaustive list of conformance tests is too long to provide here; the reader is instead referred to the IEEE 802.11 PICS tables for information from which such a list can be compiled. Summaries of the different conformance test categories will be supplied instead.

The following summarizes the areas at which PHY layer conformance tests are aimed:

1. channel center frequencies (these usually depend on a country-specific regulatory authority),

2. receiver sensitivity and channel carrier assessment (CCA) capabilities,

3. transmitter power levels and signal quality,

4. baseband functions, modulation/demodulation details, and PHY bit rate support,

5. format of the PHY Layer Convergence Protocol (PLCP) frame.

MAC-layer conformance tests are grouped into the following categories:

1. Distributed Coordination Function (DCF) and Point Coordination Function (PCF) mechanism and timing.

2. Support for MAC-layer protocol handshakes and mandatory capabilities.

3. MAC-layer security functions.

4. Wireless medium-related capabilities: rate adaptation, scanning, and synchronization.

5. Format of the different types of MAC frames.

6. MAC-layer addressing.

Each category generally consists of several individual tests. Thus for example, the MAC frame format category would include tests that exhaustively verified that each of the 25 different MAC frame formats could be transmitted and received correctly. It is easy to see that a full 802.11 conformance test suite will include hundreds if not thousands of conformance tests.

The development and implementation of each individual conformance test is fairly involved and specialized. As a typical example, consider a simple conformance test to verify that an AP (acting as the DUT) correctly performs the duplicate frame detection and recovery function required by the IEEE 802.11 PICS entry PC3.11, and defined by subclause 9.2.9.

The first step is to set up a wireless LAN test client that associates with the DUT and transfers data to it. This test client is usually created using a conformance tester such as the VeriWave WT-1210. However, it may also be produced using a standard laptop with a wireless WLAN client adapter card; custom software must then be used in place of the device driver to enable arbitrary frame sequences to be transmitted.

Once the client is set up to transfer test data frames to the wireless port of the DUT, the response data frames emerging from the Ethernet side of the DUT are checked. Checking can be done with an Ethernet packet sniffer, for example. Each unique test data frame should result in one (and only one) response data frame being forwarded on the Ethernet side. The exact type of test data frame used is not relevant; most tests use some form of UDP/IP encapsulation of an arbitrary payload, which is then converted to an Ethernet frame by the DUT and passed on.

The actual conformance test is then carried out by transmitting two identical MAC frames with the same 802.11 sequence number to the DUT. The DUT is considered to be conformant (i.e., the test passes) if the second, identical MAC frame is dropped by the DUT, without being forwarded to the Ethernet side. Otherwise, the test fails and the DUT is non-conformant.

A complete conformance test suite for an 802.11 device comprises several hundred such tests. It should be clear from the foregoing example that conformance testing is a rather exacting and detail-oriented science! Most manufacturers entrust institutions such as the University of New Hampshire Interoperability Lab (UNH-IOL) or the Wi-Fi Alliance with conformance testing of their products, rather than attempting to develop the test suites themselves.

5.2.4 Functional Testing

As previously noted, functional tests are performed to verify that a product (device or system) adheres to its stated requirements and lives up to its datasheet. The tests used are not unlike conformance tests, in that they look for specific DUT behavior under carefully controlled circumstances, but they are not quite as rigorous. It is also common for a single test to be used to verify several functions of a product at once, rather than building different tests for each function.

To take a concrete example, consider the task of verifying the frame forwarding behavior of an AP, that is, the exchange of frames between the wireless and Ethernet ports of the device. The vendor may advertise that the AP supports all frame sizes ranging from 64 to 1518 bytes (referenced to the Ethernet side), and that it can generate wireless frames with WEP, TKIP or AES-CCMP encryption, or none at all. The vendor's QA department is then tasked with proving that the AP in fact supports all this. How would they go about it?

Assuming that the appropriate test equipment is available, a single functional test is usually devised to verify this behavior. (As this is a relatively basic test, it is also normal for such a test to be included in an automated test suite when it is completed, so that it can be run on every new revision of the AP firmware.) The test is set up as follows.

Firstly, the DUT (AP) is placed in an isolation chamber and cabled to a suitable piece of WLAN test equipment (such as the VeriWave WT-90). The test equipment should be capable of injecting test traffic of different types into either the WLAN port or the Ethernet port of the DUT (or both), and measuring the response in terms of the number of packets forwarded.

The test involves a large number of individual frame sizes (the range from 64 to 1518 bytes, or 1455 unique sizes), so some automated means of sweeping over this range is highly desirable. If test equipment such as the WT-90 is employed, a frame size sweep can be used; otherwise, a suitable script (written in a language such as Tcl or Perl) will have to be written to run a test at each individual frame size and record the results.

Secondly, there are four distinct encryption modes (none, WEP, TKIP, and AES-CCMP). Again, if the test equipment and DUT permit, the four modes can be set up and run in turn without operator involvement. However, most WLAN APs require manual intervention to switch them from one encryption mode to another, so it is possible that the four modes will be tested in sequence (i.e., treated as four separate test runs). This source of manual overhead can be avoided if the DUT itself can be configured from the same script that drives the tester.

Finally, the test must be performed with traffic flowing in the wireless-to-Ethernet, the Ethernet-to-wireless, and both directions. There are hence a very large number of combinations ($1455 \times 4 \times 3 = 17460$ in all). After all of this has been comprehended, the sheer number of combinations will usually cause the QA department to realize that this is an excellent candidate for a scripted, automated test.

The script that is created should perform a traffic forwarding test at each of the combinations of test conditions (referred to as a trial). Basically, the script sets up the test equipment to generate traffic having the desired frame size, encryption mode, and traffic direction, and then starts the traffic flowing; after a specific trial duration or a pre-set number of transmitted frames, the script stops the traffic and measures the difference between the number of frames transmitted to the DUT and the number of frames received from the DUT, which should be zero if the trial has succeeded. Once the trial has completed, a new set of test conditions – frame size, encryption mode, and traffic direction – is selected and the equipment set up accordingly, and the next trial is run.

A single functional test like the above can take quite a long time to complete, even if it is scripted. However, by the same token it also exercises a great deal of DUT functionality in a single test, and can therefore expose many bugs and issues before the product is released. Many such functional tests are grouped to create a vendor's QA test plan, which is run on every product version prior to making it available to customers. As can be imagined, the time taken to complete a test plan for a product such as an AP can range from several weeks to several months, depending on the complexity of the product and the level of quality required.

5.3 Interoperability Testing

Interoperability testing is diametrically opposed to both functional and performance testing. The latter is normally carried out with specialized test equipment, and tests the DUT in isolation; that is, the test equipment strives as far as possible to avoid affecting either the test results or the behavior of the DUT. Interoperability testing, however, is done using the actual peer devices that the DUT will work with in the customer's environment. For example, an AP manufacturer would carry out interoperability testing with all of the client adapters (or at least as many as possible) that are expected to interwork with his or her AP. Conversely, a client chipset vendor would try to test their chipset reference design against as many different commercially available APs as made sense.

Obviously, both the DUT and the peer device will affect the test results; in fact, sometimes the behavior of the peer device can have a greater impact on the results than the DUT itself. Hence, the results of an interoperability test performed against one peer device frequently do not have any relationship to the results of the identical interoperability test performed against another peer device. Thus, when quoting the results of an interoperability test, it is necessary to describe not only the DUT but also the peer device against which it is tested.

5.3.1 Why Test Interoperability?

One might expect that a sufficiently detailed set of functional and conformance tests would be enough to fully qualify a DUT. After all, if the DUT completely conforms to the IEEE 802.11 standard, and also fills all the requirements of the datasheet, why is it necessary to test it against other commercial devices?

Much of the wired LAN world in fact does operate this way; instead of performing ever more exhaustive interoperability tests on ever increasing combinations of Ethernet devices, the Ethernet industry simply requires that all vendors verify compliance to the Ethernet (IEEE 802.3) standard. As the Ethernet standard is simple and well-understood, it is relatively straightforward to guarantee interoperability in this fashion. The fact that Ethernet "just works" is a testament to a single-minded focus on interoperability and simplicity by the IEEE 802.3 standards committee.

The WLAN protocol suite, however, is a different animal altogether. The WLAN MAC and security protocols are quite a bit more complex than Ethernet, and contain a very large number of moving parts. (As a comparison, the 802.3 standard requires just 66 pages to fully specify the Ethernet MAC, but the corresponding portion of the 802.11 standard, including security and Quality of Service (QoS), occupies more than 400!) Further, there is considerable latitude for implementers to introduce their own "creativity" and – in some cases – misinterpretations. All of this conspires against interoperability.

Wireless LAN manufacturers therefore have little choice but to verify interoperability by experiment. All vendors maintain large collections of peer devices – client adapters in the case of AP vendors, APs in the case of client vendors – against which each of their products is extensively tested.

5.3.2 Interoperability vs. Performance

It has been noted previously that functional tests are often confused for performance tests and vice versa. In the same vein, interoperability tests sometimes masquerade as performance tests, particularly when the test results are quantitative measures of things like traffic forwarding rate that are of intense interest to the marketing department.

One of the most crucial requirements of a successful performance test is that the test equipment and test setup should have as little influence as possible on the measured results. (One could regard this as a sort of Heisenberg principle of testing; if tester imperfections affect the measurement to a perceptible degree, then it is unclear as to whether it is the DUT or the test equipment that is being tested.) Designers of commercial protocol test equipment go to extreme lengths to ensure that their equipment is as close to being "invisible" to the DUT as permitted by the protocol standard. The test results are hence valid in the absolute sense, in that they represent a physical property of the DUT as a stand-alone device.

This is certainly not the case for interoperability tests, which have a quite different underlying philosophy and purpose. A traffic forwarding rate test performed between a specific AP and a specific client produces a result that is only valid for that particular combination of AP and client; substituting a different client but keeping the AP the same, is highly unlikely to produce the same result. Therefore, any test involving a "real" AP and a "real" client must be regarded as an interoperability test, and the results should be treated as being valid for only that combination. It is a mistake to assume that the results are valid for the AP in isolation; interoperability tests should be used only in a relative sense.

5.3.3 The Wi-Fi® Alliance Interoperability Tests

Interoperability tests are not the sole province of WLAN equipment vendors; the Wi-Fi® Alliance, an industry marketing and certification group, maintains and performs a large set of what are basically interoperability tests before certifying a WLAN chipset or device as "Wi-Fi® certified". A standardized set of four reference clients is used to test AP devices, and another standardized set of four reference APs is used for clients.

Wi-Fi® Alliance interoperability tests cover many different areas: basic performance, security (WPA and WPA2), QoS (WMM and WMM-SA), Voice over IP (VoIP), hot-spots, etc. Most of them are intended to be conducted with a low cost and fairly standardized test setup, shown in the figure below.

For each subject area, the Wi-Fi® Alliance builds a set of compliance and interoperability test procedures around the above setup, in order to verify basic standards compatibility of the equipment as well as to determine if one vendor's equipment will communicate with another.

Originally, the Wi-Fi® Alliance was concerned solely with ensuring equipment interoperability. (In fact, it was formerly known as the Wireless Ethernet Compatibility Alliance, or WECA, reflecting this intended role.) However, as 802.11 WLANs have increased in both market size and capability, the Wi-Fi® Alliance has correspondingly expanded its charter. It now includes such activities as marketing and "802.11 public relations", defining profiles – subsets of planned or current 802.11 MAC protocol functionality – to be used by early implementations, and sometimes even creating new 802.11 MAC protocols in advance of the actual standardization by the IEEE.

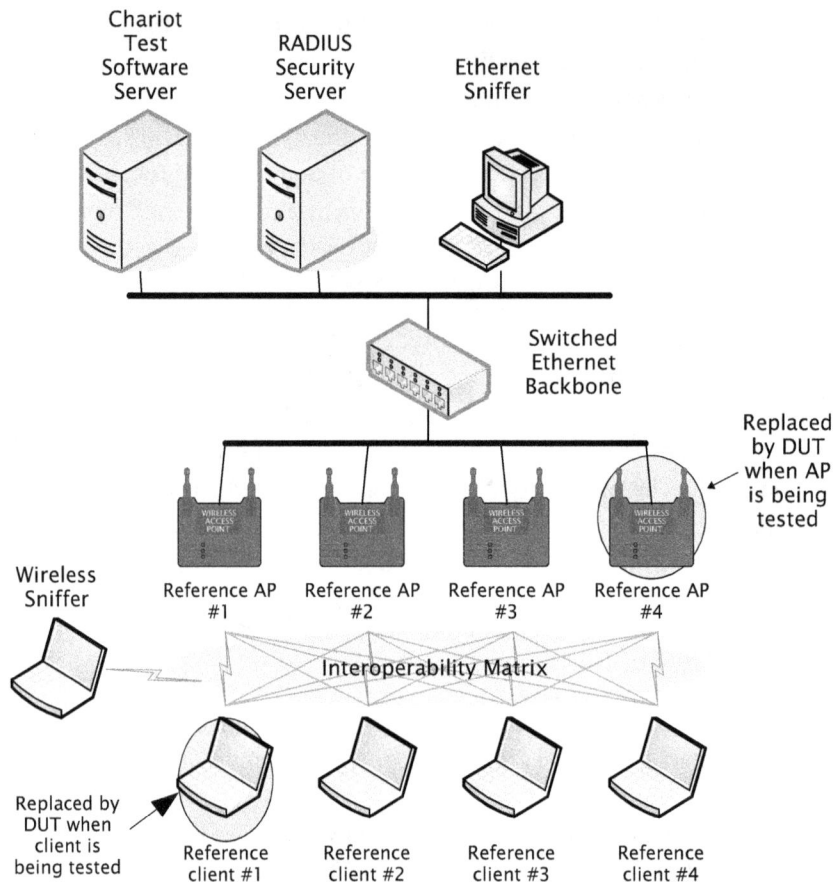

Figure 5.3: Wi-Fi® Alliance Interoperability Test Setup

5.3.4 The Interoperability Test Process

Unlike functional or performance tests, interoperability testing involves two DUTs – in the case of WLANs, this is typically an AP and a client device such as a laptop or handheld. It is not particularly useful to single out either one of the devices. Instead, the pair of DUTs is treated as a single system under test.

Apart from the DUTs, the test setup requires some means of generating traffic, and some means of analyzing it. In the case of the Wi-Fi® Alliance tests, a software traffic generator such as Chariot is used, along with a software traffic analyzer such as AirMagnet or Airopeek. In addition, an isolated environment (a conducted setup, a chamber or a screened room) is strongly recommended for repeatability's sake. However, as the test results are not particularly precise, engineers performing interoperability tests frequently conduct them in the open air.

The test process is quite simple. The two DUTs are configured into the desired mode, associated with each other, and then the traffic generator is used to inject a stream of test traffic into one of the DUTs. The sniffer or traffic analyzer captures the resulting output traffic from the other DUT. A post-analysis phase yields test results, such as the packet forwarding rate of the DUT combination, that indicate interoperability. If the test is performed with several combinations of DUTs (as in the case of the Wi-Fi® Alliance test suites) then a large matrix of results is produced, representing the different combinations.

5.4 Performance Testing

Performance testing is valued more highly than any other form of measurement within the networking industry (and in others as well, notably microprocessor technology). Most LAN users are not particularly concerned about whether their equipment meets the standard as long as it generally works, but almost all of them are deeply interested in how *well* it works. Performance tests are designed to answer that question.

The performance of networking gear may be measured along many different axes; some are purely objective (such as "throughput"), while others are subjective (such as "manageability"). For obvious reasons, this book will concern itself only with objective performance metrics. Examples of such metrics are throughput, latency, client capacity, roaming delay, etc.

In addition, note that PHY layer performance has a significant impact on the perceived capabilities of equipment. As an example, a WLAN client adapter with a significantly better radio will obviously provide a higher range than other adapters. However, these metrics have already been dealt in Chapter 4 (see Section 4.5). This chapter focuses on performance metrics that are relevant to the MAC and other packet processing layers.

5.4.1 Performance Test Setups

Performance test setups are not unlike functional test setups, in that in their simplest form they can be reduced to two elements: a DUT and a tester. However, users rarely have the luxury of confining themselves to something so uncomplicated. Actual performance test setups involve additional equipment, particularly if the tests are carried out in the open air.

Open air performance test setups involve little in the way of RF plumbing, and are closest to the normal "use model" (i.e., how a consumer or end-user would use the DUT) and hence are quite widely used. However, the caveats in Section 3.3 apply, and should be religiously observed. It is very easy for a poorly managed open-air test to produce utterly useless results when unsuspected interference or adjacent WLANs are present. At a minimum, the use of a good wireless "sniffer" to detect co-located networks and form a rough estimate of the local noise level is mandatory. Adding a spectrum analyzer to the mix is also highly recommended;

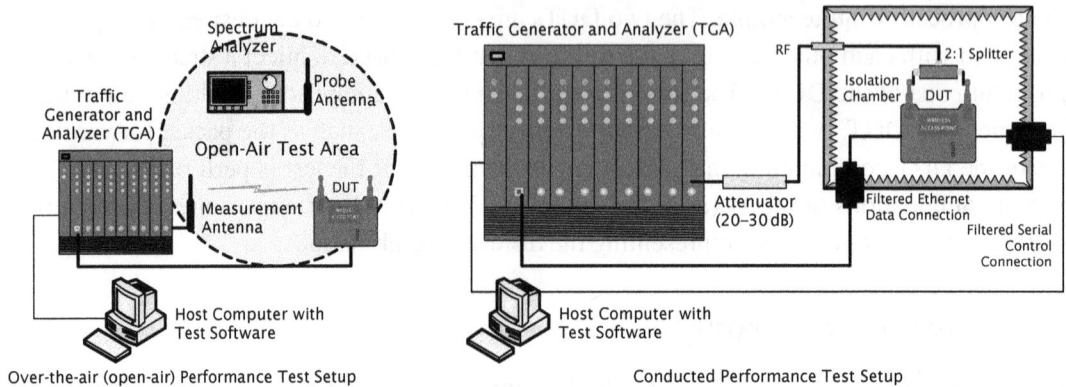

Figure 5.4: Typical Performance Test Setups

a spectrum analyzer can detect and display non-coherent interference sources (e.g., the proverbial microwave oven) that can seriously affect the test results.

One factor in open-air testing that is often overlooked is the need to account for the antenna radiation patterns of both the DUT and the tester. For example, the DUT (particularly if it is an AP) may be equipped with a supposedly omnidirectional vertical antenna; however, unless this antenna is located in the center of a large flat horizontal sheet of metal, it is unlikely to have an omnidirectional radiation pattern. (A laptop or handheld does not even pretend to have an omnidirectional pattern!) Some directions will therefore provide higher gain than others; a side-mounted or rear-mounted vertical antenna can have up to 10 dB of variation between minimum and maximum gain directions (i.e., front-to-back or front-to-side ratio). Further, coupling to power or network cables can produce further lobes in the radiation pattern in all three dimensions. What this effectively translates to is that rotating the DUT or the tester even slightly, or translating it in the horizontal or vertical directions, can cause relatively large variations in signal strength and hence affect the performance.

To eliminate the antenna radiation patterns as a source of uncertainty (at least in the horizontal direction), turntables are used. The DUT and tester are placed on turntables and rotated in small steps, typically 10° or 15° increments. The performance measurements are repeated at each step, and the final result is expressed as the average of all the measurements. This leads to a much more repeatable measurement result.

Conducted test setups are more complex and require RF plumbing. The key factors here are obtaining adequate isolation, while at the same time ensuring that the right levels of RF signals are fed to the various devices. Frequently, more than just the DUT and the tester are involved; for example, additional sniffers to capture and analyze wireless traffic, power meters to determine the exact transmit signal levels from the DUT, vector signal analyzers to measure the signal quality from the DUT, variable attenuators for signal level adjustment, and so on, may be used in the same test setup. Fixed attenuators, high-quality RF cables, properly terminated splitters, and good shielded enclosures are all essential components of a conducted test setup. The reader is referred to Section 3.5 for more details.

5.4.2 Goals of Performance Testing

Performance tests, if carried out properly, seek to quantify certain specific performance metrics. The goal of a performance test is therefore to use a test plan to measure and report a metric. It should thus be obvious that there are two essential components to every performance test: a well-defined metric, and a well-executed test plan to quantify that metric. Missing one or the other generally results in a lot of wasted effort.

A "metric" refers to the quantity or characteristic that is being measured. More time and energy is wasted on a poorly defined metric than on any other cause. The importance of knowing exactly what is to be measured, and being able to describe all of the test conditions that must be set up before measuring it, cannot be overstated. An additional requirement that is often overlooked is the need to form an abstract "model" of how the DUT affects the metric being measured. Without such a model, the test conditions cannot be properly defined, and the final measurement cannot be sanity-checked.

To take a specific example, consider the problem of measuring the latency through the DUT. The first task is defining "latency". On the face of it, this seems quite simple: latency is merely the delay through the DUT – basically, transmit a packet to the DUT and have the tester measure the time taken before the same packet is received.

However, there are some issues. Firstly, a packet contains a number of data bits, and thus takes a finite amount of time to transfer. Do we measure the delay starting from the first bit of the transmitted packet, or the last bit? The same dilemma applies to the received packet. In fact, there are four measurements possible: first transmitted bit to first received bit, first transmitted bit to last received bit, last transmitted bit to first received bit, and last transmitted bit to last received bit. Which one is correct?

For this, we turn to RFC 1242 (which deals with benchmarking terminology), which states that for store-and-forward devices, which includes WLAN equipment, the latency is measured from the last bit of the transmitted frame to the first bit of the received frame.

Another important question to answer for the latency measurement is: at what rate should we transmit packets to the device? If the rate of transmission is low (e.g., 1 packet per second), then the DUT may turn in an artificially low latency number; after all, real networks do not have such low traffic loads. On the other hand, if the packet rate is too high, then the internal buffers in the DUT will fill up (and possibly overflow), in which case we will wind up measuring the buffer occupancy delays in the DUT, not the intrinsic packet forwarding delay. The proper selection of a traffic load level for a latency test is thus quite significant. Typically, a throughput test is performed on the DUT, and the traffic load is set to 50% to 90% of the measured throughput.

Clearly, even the simplest tests can involve a number of factors, along with some knowledge of how the DUT is constructed. Forming a model of the DUT and applying it to the metric in order to properly specify and control the test conditions is one of the key challenges of performance measurement.

Once the metric has been defined and the test conditions have been specified, the next issue is creating a suitable test plan. In essence, the performance test plan is a recipe – it specifies the equipment that will be used, shows how the equipment will be connected, defines the various settings for both the DUT and the test equipment, and then gives the procedure for actually conducting the test. (Most good test plans, like most good recipes, also contain instructions for how to present the results.) It is important to write all this down and follow it exactly, in order to produce repeatable results.

A little-understood detail in the process of constructing a performance test plan is quantifying the error bounds for the test. The error bounds basically define the uncertainty of the test results; for instance, if the error bounds for a latency test were $+/-5\%$, then a measured latency value of $100\,\mu s$ could be as much as $+/-5\,\mu s$ in error (i.e., the true latency value could be anywhere between 95 and $105\,\mu s$). Error bounds are very useful in determining if the test results are valid; for example, if the calculated error bounds for a test are $+/-5\%$, but the actual run-to-run measurements vary by $+/-20\%$, then clearly something is wrong. The process of determining the error bounds for most protocol performance tests is, unfortunately rather cumbersome, especially due to the complex interactions involved between DUT and tester. Nevertheless, an effort should be made to quantify it, if at all possible.

5.4.3 Performance Test Categories

Protocol-level performance tests can be generally categorized into three types: rate-based metrics, time-based metrics, and capacity metrics. All three types are of interest when measuring the performance of WLAN devices.

Rate-based metrics measure parameters such as throughput, that are essentially the rates at which events occur. Time-based metrics, on the other hand, measure in terms of time intervals; packet latency is an example of a time-based metric. Capacity metrics, along various dimensions, measure amounts; for example, the buffer capacity of a WLAN switch measures the number of packets that the switch can store up during congestion situations before it is forced to drop one.

5.4.4 Rate-based Metrics

Rate-based metrics are the most well-known performance metrics, as they relate directly to things such as network bandwidth that interest end-users the most. Rate-based metrics include:

- throughput,
- forwarding rate (both unicast and multicast),
- frame loss rate,
- association rate.

The difference between throughput and forwarding rate is subtle and often mistaken (or misrepresented!). The best definition of "throughput" may be found in RFC 1242: to quote, it is the maximum traffic rate at which none of the offered frames are dropped by the device. Thus the frame loss rate must be zero when the traffic rate is equal to the throughput. Forwarding rate, on the other hand, as defined in RFC 2285, does not have the "zero loss" requirement; the forwarding rate is merely the number of frames per second that the device is observed to successfully forward, irrespective of the number of frames that it dropped (i.e., did not successfully forward) in the process. A variant of this metric is the maximum forwarding rate, which is the highest forwarding rate which can be measured for the device. The basic difference between whether a metric represents throughput or represents forwarding rate therefore lies in whether frames were dropped or not.

Another source of confusion in rate-based testing stems from the terms "intended load" and "offered load". The intended load is the traffic rate that the tester tried to present to the DUT; typically, this is the traffic rate that the user requested, or that the test application configured. Offered load, however, is the packet rate that the tester was actually able to transmit to the DUT. The offered load can never be greater than the intended load, but may be less. If the tester is functioning properly, a lower offered load results only from physical medium limits – that is, the PHY layer is simply not capable of transmitting any more packets than the measured offered load.

Throughput, forwarding rate and frame loss rate are common metrics, applicable to both wired and wireless DUTs. Association rate, however, is specific to WLAN DUTs; it measures the rate at which one or more clients can associate with an AP.

5.4.5 Time-based Metrics

Time-based metrics are less significant for data applications (after all, few people are concerned with whether it takes 1 or 2 ms to download an e-mail message), but are far more significant for voice and video traffic. In fact, for voice traffic, the level of bandwidth is relatively unimportant (as a voice call occupies only a fraction of the 20 Mb/s capacity of a

WLAN link), but the delay and jitter of the traffic has a huge impact on the perceived quality of the call.

Time-based metrics include, among many others:

- latency,

- jitter,

- reassociation time.

As WLAN APs and switches are universally store-and-forward devices, latency is normally measured from the last bit of the frame transmitted to the DUT to the first bit of the corresponding frame received from the DUT (see Section 5.4.2). It is typical to measure latencies by specially marking individual packets in the transmitted traffic from the tester (referred to as timestamping the traffic, and often accomplished by inserting a proprietary signature into the packet payloads containing identification fields) and then measuring the time difference between transmit and receive. It is common to average the measured latency over a large number of packets in order to obtain a better estimate of the DUT performance.

Jitter is a measure of the variation in the latency, and is of great interest for real-time traffic such as video and voice. Different jitter metrics have been defined: peak-to-peak jitter, RMS jitter, interarrival jitter, etc. The jitter metric commonly used in LAN testing is defined in RFC 3550 (the Real Time Protocol specification), and is referred to as smoothed interarrival jitter. It is, in essence, the variation in delay from packet to packet, averaged over a small window of time (16 packet arrivals).

Reassociation time is unique to WLANs; this is the time required for a WLAN client to reassociate with an AP after it has disconnected (or disassociated) from it, or from another AP. This is important measure of the time required for a network to recover from a catastrophic event (e.g., the loss of an AP, requiring that all clients switch over to a backup AP).

5.4.6 Capacity Metrics

Capacity metrics deal with amounts, and are mostly applicable only to WLAN infrastructure devices such as switches and APs. A classical example of a capacity metric is the association database capacity of an AP. APs need to maintain connection state for all of the clients that are associated with them; the upper bound on the number of clients that can connect to the same AP is therefore set by its association database capacity. (Of course, other factors such as bandwidth and packet loss also kick in when sizeable amounts of traffic are generated.)

Other capacity metrics include burst capacity and power-save buffer capacity. Burst capacity is the ability of an AP or switch to accept back-to-back bursts of frames and buffer them up for subsequent transmission; this is important because LAN traffic exhibits highly bursty characteristics, and it is essential to be able to deal with traffic bursts and ensure that frames are not lost. Power-save buffer capacity is a similar metric, measuring the APs ability to buffer frames destined for its clients, but in this case involves WLAN clients that are alternating between sleep and wake modes and thus have periods where they cannot accept traffic. (Sleep mode is used for conserving battery life in laptops, handsets, and Personal Digital Assistants (PDAs).)

5.4.7 Scalability Testing

Large enterprises usually require correspondingly large WLAN installations. For instance, a typical large office building might serve a thousand users, using two to three hundred APs and perhaps a half-dozen WLAN switches. Both the IT staff and their equipment vendors are interested in ensuring that such a large network stays up and performs well under all kinds of traffic loads, sometimes on a 24-hour-a-day, 7-days-a-week basis. A variation of performance testing is performed to ensure this, falling under the category of scalability testing.

Scalability testing basically involves building out a substantial network topology – in many cases using exactly the same devices that are expected to be deployed in the actual "production" network – and subjecting it to a variety of performance and stress tests. For example, a network vendor, in order to prove to a customer that their equipment will not fall down, might create a full-blown LAN: 50–100 APs, 5–6 WLAN switches, 2–3 Ethernet routers, DHCP servers, Remote Authentication Dial-In User Service (RADIUS) authentication servers, file servers, WAN gateways, VoIP call servers, etc. The LAN is then subjected to simulated traffic to exercise its functionality and capacity. (Obviously, packing a thousand users into a building in order to test the LAN with "live" traffic is not very feasible!) Various performance tests – throughput, latency, client capacity – are run on the entire LAN, to verify that the scaled-up LAN works as well as its individual components.

Scalability testing gets particularly important when WLANs are applied to "mission critical" applications. For example, having a WLAN in the lobby go down is merely a nuisance. However, if a WLAN is serving the corporate phone system and carrying traffic from VoIP/WLAN handsets, it becomes a disaster. Scalability testing is essential for avoiding such disasters.

It is usually not necessary to reserve an entire building for scalability tests (i.e., in order to physically deploy the APs and the traffic simulators in an over-the-air environment). Over-the-air scalability tests become rapidly more expensive and impractical as the size of the network is increased. Instead, scalability testing is most easily performed with a fully cabled setup. As the performance of the infrastructure is being measured, rather than the behavior of the RF channel or the coverage of the APs, there is very little loss in terms of "realism" or

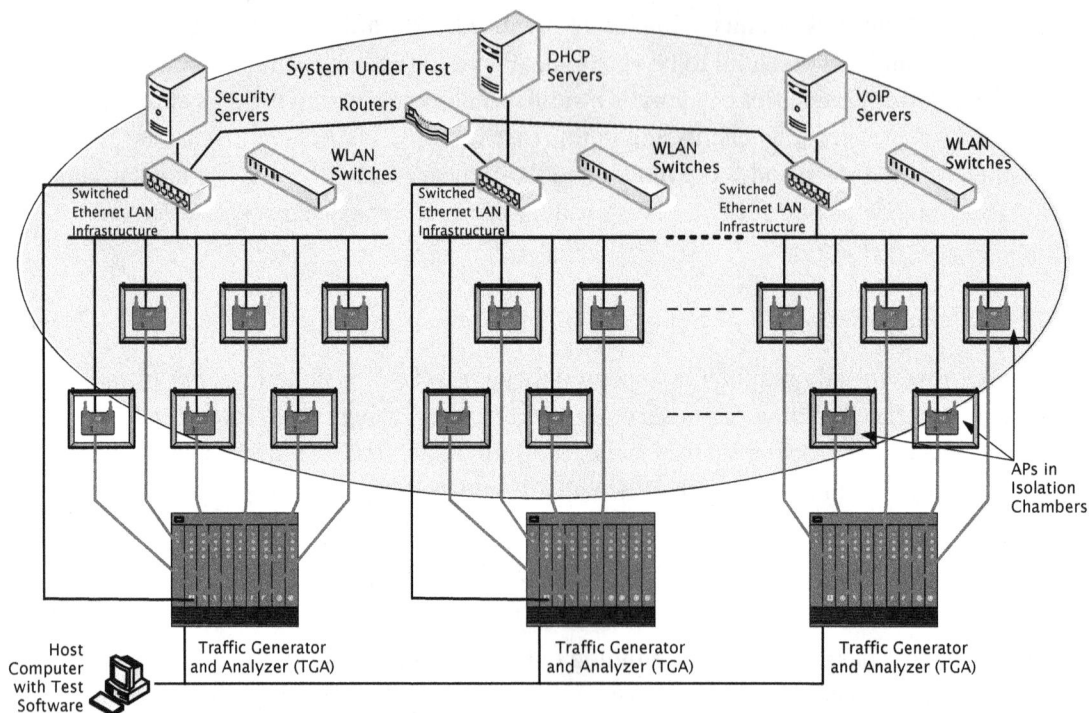

Figure 5.5: Large-Scale Performance Testing

network stress. Obviously, realizing the measured performance in actual practice is dependent on proper deployment installation practices (covered in a later section), but with modern equipment and good site planning this is no longer an issue.

5.4.8 Roaming performance

A unique characteristic of WLANs (and, in fact, one that endears them to users in the first place) is that their users are not static, tethered to desks by cables. Instead, they move about from place to place. Mobility testing in various forms is a key component to quantifying the performance of WLAN equipment.

A clarifying note is necessary here, for the benefit of those familiar with terminology in the cellular industry. The WLAN industry refers to the process of clients moving from one AP to another in the same LAN as "roaming". However, this is equivalent to cellular handsets moving from one basestation to the next, within the network of the same service provider – a process which the cellular industry refers to as "handoff" or "handover". Roaming, as used in the cellular industry, refers instead to the process of connecting to the network of a different service provider – i.e. a different network altogether. Roaming in WLANs is therefore equivalent to handover in cellular networks.

Roaming performance is characterized by several metrics:

- average roaming delay,
- packet loss during roaming,
- dropped connections.

The average roaming delay is simply the time required for the client to disconnect from one of the APs and reconnect to another one. The roaming delay is typically measured in terms of the arrival times of specially marked or timestamped data packets, so as to establish exactly when the client disconnected or reconnected. Roaming delays can range from the small (10–20 ms) to quite large values (several hundred milliseconds) depending on the security mode employed. In fact, a roaming test is a fairly strenuous test of the security infrastructure.

Measuring the packet loss during roaming is of interest because it quantifies the DUT's ability to hide roaming artifacts from upper-layer (i.e., TCP/IP) protocols. A typical roaming delay of 50 ms can represent a considerable number of TCP segments, if the roaming process occurs in the middle of a file transfer. The ability of the client or the WLAN switch to hide the delay and prevent packet loss from triggering large-scale retransmissions and slowdowns at the TCP layer is fairly important to good end-user experience.

The most catastrophic outcome of a botched roaming process (at least, from the user's point of view) is a dropped connection or failed roam. This manifests itself as the complete failure to resume normal data transfer after reconnecting to the new AP, and is indicative of either system overload or some deep-rooted bug. The end result of failed roams is an abruptly terminated voice call or e-mail download, so such events should be recorded and reported carefully.

Either clients or APs can serve as the DUT; if the roaming performance of the client is to be measured, only one client is required, but if the roaming performance of an AP is to be measured, then two (or more) APs of the same make and model are needed. Typically these APs will be connected to a WLAN switch of some sort, so in reality the roaming performance of the entire setup – APs and switch – will be measured as a unit.

The basic test setup for roaming measurements comprises a DUT (a client, or two or more APs, as mentioned above) together with a tester that serves as the counterpart device, as shown in the following figure.

Different methods of inducing the roaming event exist in the industry today. One approach is to use a pair of variable attenuators, adjusted so as to shut off one AP and enable the other AP; the client will automatically roam from the first to the second APs, at which point the roaming metrics can be measured. The same setup is generally used for measurements on both clients

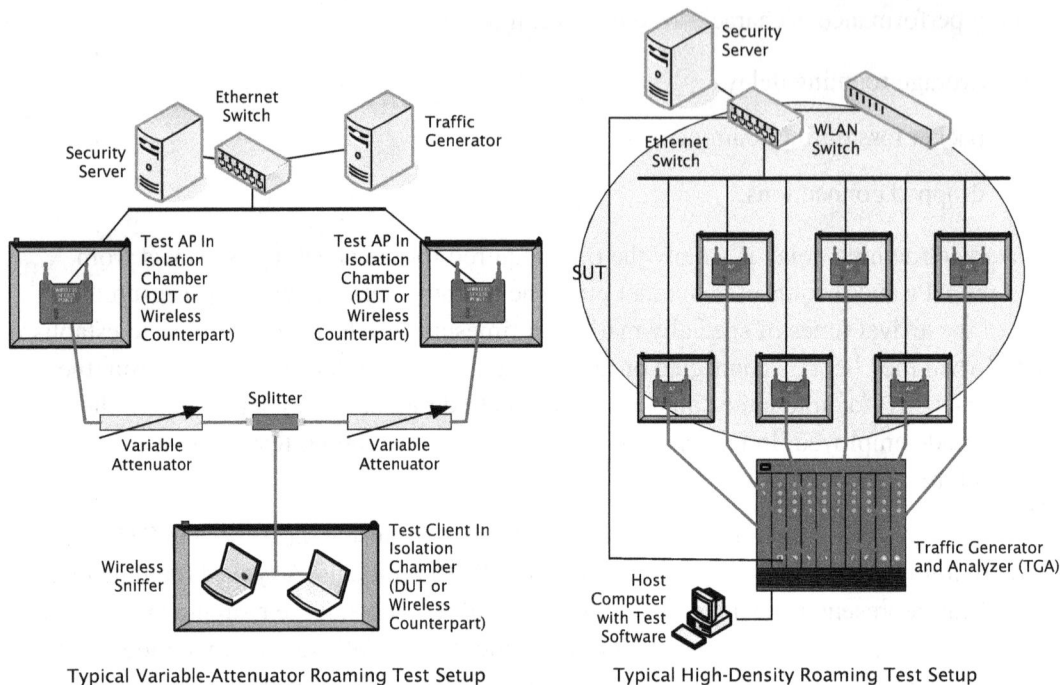

Typical Variable-Attenuator Roaming Test Setup   Typical High-Density Roaming Test Setup

Figure 5.6: Roaming Test Setup

and APs, the choice of the DUT being a matter of viewpoint rather than setup. As neither APs nor clients are designed to be test equipment, the actual roaming measurements are made by post-processing packet captures ("sniffer traces") obtained from nearby wireless sniffers during the roaming process.

Another approach, which is used for roaming measurements on APs and WLAN switches when the clients are emulated and hence are under the control of the test system, is to programmatically cause the emulated clients to roam, and make measurements during the process. This avoids the repeatability and uncertainty issues caused by the use of off-the-shelf clients (many of which have severe problems with stability and controllability), and allows the roaming test to be performed rapidly and automatically. In addition, the need for wireless sniffers or post-processing of sniffer traces is completely eliminated, and much higher-density test setups are possible.

5.4.9 QoS Metrics

The IEEE 802.11e QoS standard is widely adopted for supporting VoIP traffic over WLANs. (Actually, the subset defined by the Wi-Fi® Alliance – Wi-Fi® Multimedia or WMM – is the more common implementation, as the full IEEE 802.11e standard far exceeds what is typically required for simple VoIP applications.)

QoS adds another dimension to performance testing of WLANs. In addition to standard performance tests such as throughput and latency, QoS-specific metrics that are directly aimed at quantifying how well the WLAN supports the needs of voice and video traffic streams become interesting.

Typical QoS performance metrics can be divided into two categories:

1. Low-level measurements that focus on how well an AP or client implements the 802.11e protocol, especially the prioritization of traffic.

2. Application-level measurements that try to determine the efficiency of the QoS implementation as a whole, and how it improves service quality for real-time traffic.

Low-level QoS measurements are fairly limited, confining themselves to determining whether traffic streams in various priority classes are isolated from each other, and also whether QoS mappings are preserved when transitioning between the wired and wireless media. These are done by simply generating and mixing multiple traffic streams assigned to different QoS priority levels, and determining the level at which higher priority traffic is affected by lower priority traffic. Note that QoS priority levels do not have much effect until some degree of oversubscription or congestion occurs. Therefore, a typical test setup would ensure that the aggregate offered load of all the traffic streams presented to a WLAN device exceeded the medium capacity, and then compare the loss, latency, and jitter of the high-priority streams with that of the low-priority streams. If the high-priority streams remain unaffected even though some portion of the low-priority streams are being dropped due to oversubscription, then the QoS performance of the WLAN device is good.

Application-level QoS measurements are more interesting from an end-user point of view, as end-users are rarely concerned about individual links but instead look at end-to-end performance. Application-level QoS performance tests may be divided into two types: capacity metrics and service assurance metrics.

Capacity metrics seek to determine the total capacity of a WLAN or device when carrying delay or loss sensitive traffic. For example, in a VoIP application, it is essential to determine the maximum number of concurrent voice streams that can be carried before the QoS guarantees break down; this aids the end-users (either enterprises or service providers) in provisioning their network to deal with the maximum expected load. Capacity metrics are usually measured in the presence of some fixed amount of background traffic. For example, a capacity test could inject 1 Mb/s of best-efforts data traffic into a WLAN switch, and then measure the maximum number of voice calls that could be carried by that switch before the voice quality dropped below a minimum threshold.

Service assurance metrics look at the ability of a WLAN or device to prevent delay and loss sensitive traffic from being affected by the bursts of data packets that are encountered in heavily loaded LANs. Once a voice call is established, maintaining call quality requires that the network give priority to the VoIP traffic carrying the call at the expense of best-efforts data such as e-mail and file transfers. (A separate mechanism – referred to as call admission control or CAC – is used to limit the number of voice calls to the capacity of the network.) A service assurance measurement is therefore made by setting up a pre-defined number of voice calls, and then gradually increasing the amount of best-efforts data traffic until the voice quality drops below a minimum threshold.

Determining the application-level quality of a delay or loss sensitive traffic stream is an art in itself. For voice streams, well-defined metrics exist, such as the R-factor (also called the R-value) defined in ITU-T G.107; the R-factor attempts to measure call quality in terms of the predicted satisfaction level of humans listening to the voice call. More accurate, but also more complex, perceptual speech quality metrics (PESQ and PSQM) have also been defined for quantitative measurements of VoIP traffic. Video stream quality measurement is, however, still in its infancy. To date, video quality metrics have been subjective (i.e., relying on human viewers to rate the quality of video streams) rather than objective. Organizations such as the Video Quality Experts Group (VQEG) and even some companies such as Intel and IneoQuest are attempting to define objective video quality metrics, but this is a research topic as of the writing of this book.

5.4.10 An Alternative Classification

The same performance metrics can also be classified into two categories: data-plane metrics and control-plane metrics. This classification becomes useful when trying to understand which portion of a DUT is affected by a specific metric, and, by extension, where to start looking if a test produces a poor result.

Data-plane metrics pertain to the performance of the basic frame processing and forwarding functions within the DUT (i.e., what is commonly referred to as the DUT datapath). For example, a throughput test directly measures the ability of a DUT to receive, classify, switch, buffer, and retransmit frames; low results on a throughput test usually indicate some issue in one of these areas, such as packets being lost due to a poorly selected buffer management algorithm.

Control-plane metrics apply to the state and context management functions within the DUT, such as those related to client state update when handling mobile clients. An example of a control-plane metric is roaming delay. Roaming is usually quite stressful on the control functions within the DUT, because the client context at the original location of the client must be torn down and new context established at the new location, all within as short a time as possible. The impact of the DUT forwarding path is at best second order; a DUT with poor throughput may still be able to offer superior roaming performance, provided that careful attention has been paid to the design of the control software that maintains client context and manages security.

The following table summarizes a non-exhaustive set of data-plane and control-plane metrics:

Data-Plane Metrics	Control-Plane Metrics
Throughput	Association rate
Forwarding rate	Reassociation time
Frame loss rate	Association database capacity
Latency	Roaming delay
Jitter	Dropped connections
Burst capacity	Reset recovery
Power-save buffer capacity	
QoS capacity	

5.5 Standardized Benchmark Testing

There is probably no area of protocol testing that rouses as much interest – and controversy! – as standardized benchmark measurements. Benchmarking brings together QA engineers, marketing people, magazine publishers and even end-users, all of whom are attempting to answer one question: which equipment is better? It is in fact the one time when the activities of the QA department can actually make it to the top of the CEO's priority list.

5.5.1 Lies, Damned Lies, and Benchmarks

Benchmark testing is not merely a LAN or even networking phenomenon; standardized benchmark tests are used throughout the computer industry. (For example, the Linpack benchmark originated by Jack Dongarra measures the computational performance of supercomputers.) The ostensible purpose of a benchmark test is to quantitatively rank equipment, usually from different vendors, based on some objective quality metrics. Unfortunately benchmarks, like statistics, can be structured and twisted to support any arbitrary point of view, particularly if the metric is poorly defined or the test is poorly executed (see Section 5.4.2 on performance testing).

A good benchmark test needs to bear some correlation to a non-trivial part of the end-user's experience. There is little point in testing some aspect of a networking device that may be interesting to an engineer but does not substantially affect the ultimate end-user. The nature of benchmark testing in fact indicates the maturity of an industry; initially, benchmark numbers tend to focus on whizz-bang capabilities that are of interest to technophiles, but as the industry matures the focus shifts to metrics that the end-user can actually utilize in order to better design and deploy the equipment.

Another hallmark of a good benchmark test is simplicity. It should be easy to explain to an end-user why he or she should be interested in the results of the benchmark. Not only should the benchmark itself be simple, but it should be intuitively obvious (when presented with two sets of test results) which results are better and why. The more complicated and contrived a benchmark, the less likely it is that it will have much correlation to end-user experience.

5.5.2 Comparative Testing

Virtually all benchmarks are relative in nature, even though their test results may yield absolute performance numbers. That is, benchmark test results on a given piece of equipment have little value (i.e., as benchmarks) unless they can be compared with the results of similar tests performed on other equipments of the same type. Benchmark testing is thus highly comparative in nature. In fact, most benchmark testing campaigns are performed with equipment from different vendors being tested at the same time, or at least in the same test campaign. The results are usually presented side-by-side for easy comparison.

This highlights one aspect of benchmarking, namely, the competitive nature of benchmark testing. Understandably, marketing departments usually spend a good deal of time attempting to make benchmark test results of their company's equipment look better than that of their competitors'. It is therefore essential to ensure that benchmark tests are performed the same way on all equipments; to facilitate this, a well-designed benchmark test will specify virtually all aspects of the test setup, the test conditions and the equipment configuration parameters, down to the MAC and IP addresses to be used and the time allowed between test trials.

Comparative benchmarking has become such an institution in the networking industry that there are even institutions (such as the Tolly Group, and Network Test) that have been established with the purpose of performing independent comparative tests on different vendors' equipment.

5.5.3 Typical Benchmarked Metrics

Benchmark tests are almost without exception a subset of the normal performance tests that are typically carried out by manufacturer or user QA labs on WLAN equipment or systems. The same metrics are used, and similar test procedures are followed. The primary difference is that the level of rigor, documentation, and scrutiny is much higher – QA performance tests are not usually carried out with several marketing departments breathing down the neck of the person conducting them!

The most common benchmark tests for WLAN equipment are the usual suspects:

- throughput,
- forwarding rate and packet loss,
- latency,
- scalability in terms of number of clients and APs,

- roaming measurements,

- QoS measurements.

Functional and conformance tests should obviously not be made the subject of benchmark testing. Not only are they frequently specific to the equipment, they are of little interest to an end-user.

5.5.4 Dos and Don'ts of Benchmark Testing

A properly conducted benchmark test campaign can yield results that are both useful and interesting to the end-user community. However, there are plenty of pitfalls along the way to trip up the unwary or the careless. Some of them are:

- Testing a modified DUT. Benchmark campaigns are usually accompanied by several engineers and marketing people from the companies selling the equipment, who are usually all too anxious to "tweak" the DUT in order to get better test results. In many cases these "tweaks" can render the DUT useless in a production network, though they produce excellent results when running benchmark tests. Good benchmark testers resist such pressures, and insist instead that the DUT should be set up exactly according to the instructions provided to the end-user.

- Selecting a proper benchmark traffic load. Random mixes of traffic lead to wildly varying and often unrepeatable results. Also, failure to control all aspects of the test traffic can produce quite a bit of variation, independent of DUT characteristics. The traffic injected into the DUT needs to be well-controlled and well-defined; otherwise the benchmark campaign can degenerate into a test of the patience of the person running the benchmarks.

- Recording all results. Not all equipment can score equally well on a benchmark test. In fact, if they did so, then the benchmark test would be quite uninteresting. It is not unusual for the marketing or engineering departments of the low-scoring equipment to violently challenge the quality of the test equipment, the veracity of the tester, the nature of the test traffic – everything, in fact, except their own LAN gear. (In some cases there may even be legal liabilities.) To defend against this, it is essential to save everything pertaining to each and every benchmark test until the test has passed out of the public eye.

- Misbehaving DUTs. A non-compliant DUT can make the results of performance tests very puzzling indeed. For example, a DUT that does not implement the 802.11-mandated protocol timers and inter-frame spacing can appear to have higher than possible throughput numbers. If possible, a conformance test should be conducted on the DUT before embarking on an extensive benchmark test campaign to smoke out such devices, or at least understand what could be causing strange test results.

Application-Level Measurements

Application-level measurements resonate well with end-users; they provide a sense that there is some correlation to the ultimate end-user experience, rather than being merely some abstract property of the infrastructure devices. In some situations, such as in the case of the WLAN industry of 2006 or before, the lack of more advanced test tools drives the manufacturers and vendors themselves to make application-level measurements, because such measurements can be made with little more than the actual applications themselves, or a simulacrum thereof. As a consequence, much of the WLAN industry today still relies on application-level metrics and measurements to indirectly determine the capabilities of the underlying WLAN hardware and software. This chapter covers some of the key application-level tests and setups used in current practice. The focus is on system-level testing for enterprise and multimedia applications.

6.1 System-level Measurements

Application-level measurements are also referred to as system-level measurements, because they attempt to characterize the performance of a system as a whole, rather than any piece or subset of it. Much more than simply wireless clients and access points (APs) are involved by necessity in transporting traffic from its source to its sink, particularly when dealing with real-time traffic such as voice and video. A packet voice circuit usually comprises, from start to finish: a microphone, an audio encoder (codec), a packetizer, a wireless client, an AP, a wired network infrastructure (switches and routers), a call server or Voice over IP (VoIP) gateway, another AP, another wireless client, a depacketizer, an audio decoder (another codec), and a speaker or earphone. Each element of the chain has an impact on the perceived quality of the delivered audio. Thus obtaining a true picture of the complete packet voice system necessitates end-to-end measurements, which are obviously best done at the application level.

6.1.1 Your Mileage May Vary

Application-level measurements are of immediate interest to both end-users and enterprise IT people because they *appear* to directly quantify the anticipated user experience. Making measurements of the underlying parameters of a local area network (LAN) connection such as latency and packet loss is all well and good, but nothing can match the immediacy of a direct measurement of voice quality (preferably using actual audio samples, or even human listeners). Measurements at the application layer are thus appealing to virtually all network managers and users, even technically sophisticated ones.

There is one substantial caveat, though: being end-to-end measurements, application-level metrics are affected by more than just the wireless equipment that carries the traffic. For instance, while voice quality is certainly impacted by loss and jitter in the wireless LAN, it is just as much a function of the type of codec used, the quality of the call server or gateway, and in fact the firmware (OS and drivers) running on the handsets. This necessitates a certain amount of care when using application-layer metrics to quantify the performance of WLAN equipment; poor test results do not necessarily imply that the WLAN devices are at fault, unless all other elements along the entire end-to-end path have been considered and ruled out.

A second, less well-understood problem with application-level measurements is repeatability. Almost by definition, the application layer in any protocol suite is complex and somewhat unpredictable. It is extremely difficult to set up the same initial conditions twice when running a real application, because the state of the application is determined by a myriad of software and hardware interactions as well as an operating system. Most application-level measurements rely on running many iterations of a particular test over long periods of time to average out the uncertainties in the initial state of the application layer and the random interactions with the rest of the system. In spite of this, application-layer metrics usually have to settle for a higher level of uncertainty and a lower degree of repeatability than any other kind of measurement.

Thus, one should not take small differences in measured results too seriously when conducting application-layer testing. By the same token, one should not expect more than a first-order approximation to the actual performance of the WLAN when installed in the live network.

6.1.2 Enterprise WLAN

Before plunging into how application-level measurements are performed, it is worth taking another look at the architecture of a WLAN with different types of end-user devices. Figure 6.1 depicts a simplified view of a typical enterprise WLAN setup.

A modern enterprise WLAN integrates a wide variety of equipment and technologies. The actual mix is dependent on the nature of the enterprise as well as the choices made by users and administrators; the figure only shows what is possible:

- The most ubiquitous WLAN clients are, of course, laptops. The most common data applications on these clients are e-mail, file service, and Web browsing, but other applications such as VoIP (e.g., Skype), conferencing, and video are also prevalent. Personal Digital Appliances (PDAs) are basically cut-down versions of laptops.

- VoIP handsets (basically, phones using WLAN interfaces) are also present, though less prevalent at the moment.

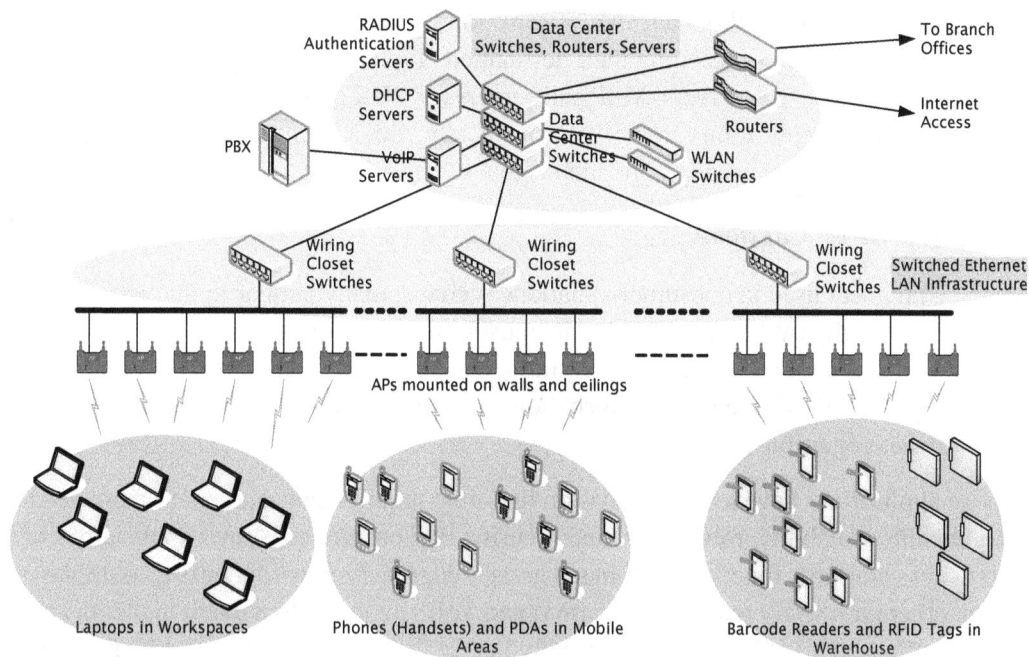

Figure 6.1: Enterprise WLAN Topology

- More esoteric devices such as bar-code readers and radio-frequency identification (RFID) tags are coming into use for asset tracking and inventory purposes.

- The above WLAN clients are provided access to the wired infrastructure by means of APs. There is usually only one wireless hop between a client and the nearest wired node; occasionally there may be more, as in the case of wireless bridges or wireless mesh networks.

- The AP connects directly to an Ethernet port on a switch; the latter is increasingly becoming a "wireless-aware" device that is capable of finding and managing APs directly.

- Various pieces of wired Ethernet gear such as switches and routers make up the rest of the wired LAN infrastructure. Most enterprises also have interfaces to an Internet Service Provider (ISP) or WAN transport links to remote sites, which require additional Ethernet or non-Ethernet equipment to support.

- Connected to the wired LAN infrastructure are several different types of servers (servers are almost never connected using wireless links – there is little use in having a server that could be picked up and moved away without warning).

All of the traffic to and from the WLAN clients invariably terminates on one or the other servers; it is extremely unusual to find packets that originate on a WLAN client and terminate on another WLAN client, without a server of some kind in the middle. Even voice packets must traverse a VoIP server before reaching their destination. Typical examples of servers are file servers, e-mail servers, Web servers, VoIP call servers (or PBX gateways), video servers, and so on. Most network applications are written specifically with a client-server model in mind, so that each client application usually corresponds to a server somewhere on the LAN or Internet.

6.1.3 Enterprise vs. Consumer

While the equipment used in consumer situations is conceptually similar to that in enterprises – after all, a WLAN reduced to its most basic form consists of an AP, a client and a wired infrastructure – the application traffic encountered is completely different. Further, enterprise WLANs tend to be fairly homogeneous, but consumer networks are often aggressively heterogeneous.

Enterprise applications have less emphasis on real-time requirements and much more need for high bandwidth, scalability, robustness, and security. By contrast, consumer applications of WLANs stress real-time functions to a much greater degree. Multimedia traffic, multiplayer games (another intensively real-time situation) and VoIP are found to a much higher degree in consumer settings, while file servers are usually not. In fact, the goal towards which consumer equipment vendors are working is a home network that enables seamless integration of home computers, entertainment devices and the Internet (see Figure 6.2).

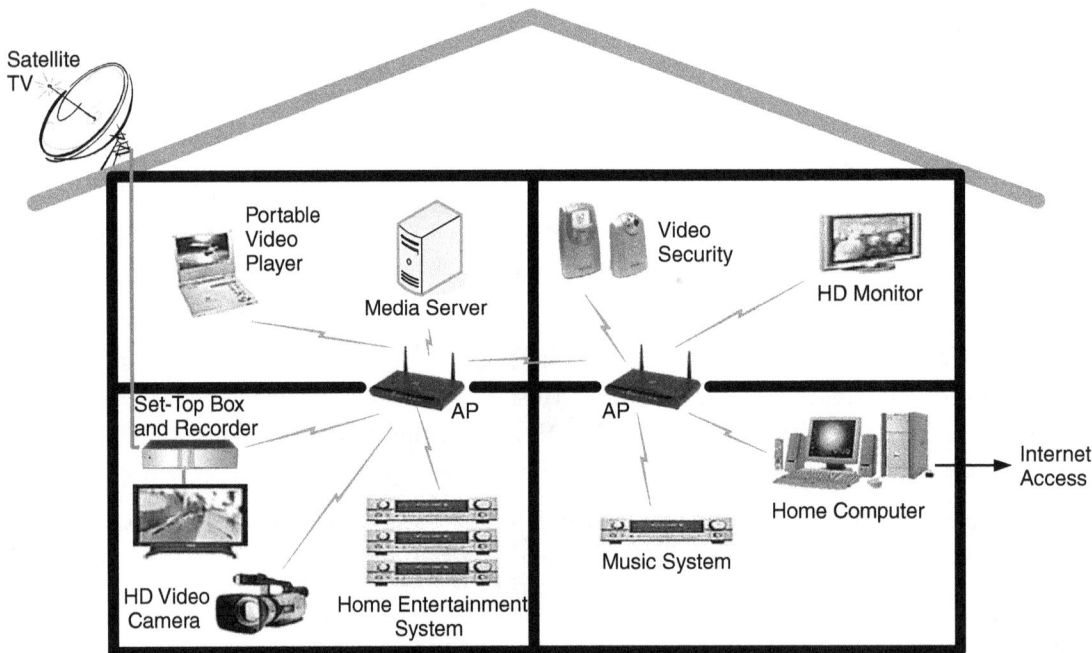

Figure 6.2: The Consumer WLAN Utopia

Traffic and test procedures that quantify the end-user experience with consumer equipment are thus quite different from those for enterprise equipment. Scalability is obviously not a concern at all – it is hard to imagine a home network with hundreds of clients – and a greater degree of uncertainty in the measurements is acceptable. Instead, coverage area, resistance to interference, support for real-time applications, and interoperability are the prime foci.

6.1.4 Vertical Applications

Wireless LANs are somewhat unique in the networking industry in terms of the rate at which they have penetrated specialized "vertical" applications. An example of a vertical application is warehousing. The use of WLANs in warehouses is common; by contrast, wired LANs have had little or no impact here. WLANs are routinely used to provide the infrastructure for bar-code readers, RFID devices, asset location and tracking systems, and automated inventory management. Other examples of vertical applications for WLANs are hospitals (everything from miniature VoIP phones to patient monitors), factories, aircraft maintenance hangars, and retail stores.

Each vertical application offers unique challenges in testing, not the least of which is generating the traffic required for driving the wireless infrastructure. For example, a WLAN-based RFID system may be required to track thousands of tags, each representing a "client"; however, testing the system by actually using such tags (and then moving them about from place to place) is clearly very expensive and not very practical. At the same time, the traffic encountered in an RFID system does not resemble that in common enterprise or consumer scenarios, so using test equipment and procedures used for the latter is not a very realistic method of testing.

Testing WLANs in vertical applications continues to be a fairly challenging problem.

6.1.5 Usage Cases

Maintaining the end-user orientation of application-level measurements, we can categorize the metrics being measured into usage cases. A usage case is an attempt to narrow the focus of a measurement scenario; each usage case represents a typical end-user activity, and is associated with a set of metrics that quantify the end-user experience. For example, all metrics related to voice calls can be grouped into a "VoIP usage case", while measurements pertaining to file transfer could be categorized in a "data usage case".

The IEEE 802.11.2 draft standard (see Appendix A) has conveniently specified three usage cases that cover virtually all end-user applications currently known to exist on a WLAN:

- A latency-sensitive usage case that encompasses applications that are highly sensitive to latency, jitter and packet loss, but much less demanding of bandwidth. The most obvious example of an application that falls into this usage case is packet voice, but network conferencing is also covered.

- A streaming-media usage case; this comprises applications that are sensitive to latency and jitter, but at the same time also bandwidth intensive. A classical example is real-time packet video, which can tolerate little latency or jitter, while still requiring megabits of sustained bandwidth from the WLAN. Most video (e.g., multicast video) and multimedia applications fall into this usage case.

- Finally, a data-oriented usage case that covers applications insensitive to delay, loss and sustained network bandwidth. Examples are Web, file transfer, and e-mail, though non-real-time video could also conceivably be placed here. Note that while the application is mostly insensitive to the sustained bandwidth provided by the WLAN, the level of user satisfaction is not; a higher bandwidth usually translates to a better user perception of the network.

Once a particular application has been categorized into a specific usage case, it is fairly simple to identify the measurements to be made on the traffic the application generates, in order to quantify the end-user experience. In fact, the IEEE 802.11.2 document cross-references usage cases against metrics and test methodologies for exactly this reason.

6.1.6 Measurement Setups

Measurement setups for application-level testing are generally more complex and more varied than any other kind of testing, due in large part to the complexity of the traffic being applied to the device under test (DUT) and the measurements that must be made. Further, it is rarely possible to test a DUT "in isolation", because of the additional devices and software needed to support the application traffic.

A typical application test setup might consist of:

1. an application-level traffic source, which may be the actual application itself, or (if possible) a more controllable and observable software simulation thereof;

2. whatever hardware interface is required to support the application and inject the traffic into the DUT; for example, this could be a laptop with a WLAN client interface;

3. the DUT, which is really an entire system (e.g., an AP, a call server, routers, switches, gateways, etc.);

4. an application-level traffic sink; this is frequently an actual server or server farm running the appropriate software, or a simulated server;

5. the wired infrastructure required for the system; and

6. traffic analysis devices to measure both the traffic stimulus applied to the DUT and the response, unless the traffic source and sink are properly instrumented for this purpose.

The following figure shows this in detail, for a voice test scenario.

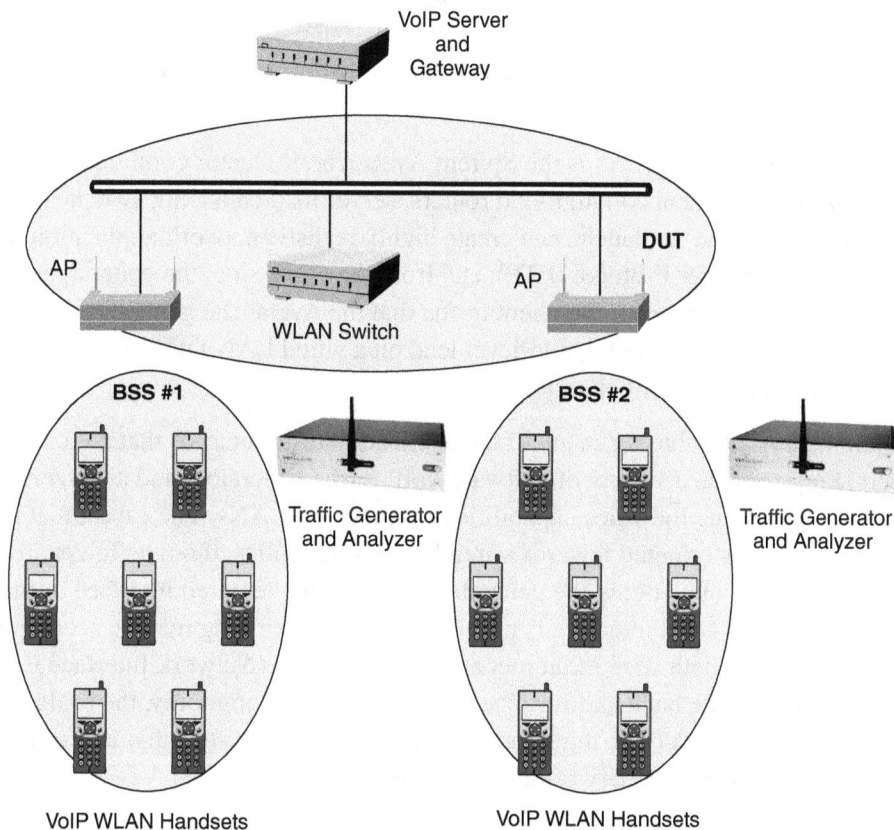

Figure 6.3: Voice Application Test Setup

Formerly, the only choice for anyone wishing to perform application-level measurements was to use the actual application as the source and sink of test traffic. For example, measuring the Web browsing performance of a system essentially involved setting up a Web server and some number of browsers on different computers connected to the system, and manually clicking through HTML documents. Obviously, the measurements available from such an approach are quite limited and crude, and the setup is labor-intensive and irreproducible.

Fortunately the "application-awareness" of the LAN infrastructure was likewise fairly limited, and hence there was little need to perform application-level tests. Tests done with network layer (i.e., Internet Protocol, or IP) traffic, or transport layer (Transmission Control Protocol (TCP)) traffic, were sufficient to exercise all of the important capabilities of the devices used.

The capabilities and complexity of even general-purpose LAN devices such as routers have grown, however, and application-aware packet processing is becoming an integral part of the LAN infrastructure. This, coupled with the increasing interest in testing performance with actual network application traffic, has sparked the development of specialized application-layer test equipment. Most such equipment is currently found only on the wired side of the world, though companies such as VeriWave are starting to develop similar devices for wireless testing.

An example of such test equipment is the Spirent Avalanche/Reflector combination, intended for performance measurements on high-end routers, server load balancing switches, gateways, firewalls, and the like. The Avalanche can create highly realistic network application traffic such as HTTP, File Transfer Protocol (FTP), etc. from the client side; the counterpart Reflector can act as an equivalent server for the client traffic that the Avalanche generates. Together, they can provide a controlled application-level load on a wired LAN DUT, and also measure the DUT performance in various situations.

Similar equipment is available for many of the application test scenarios that are of interest to end-users. For example, a variety of software VoIP traffic generators and analyzers have been created for analyzing the voice capabilities of enterprise LANs. The caveat that most of these tools are primarily oriented towards wired LANs still applies, though. To get around this limitation, some means of bridging the generated traffic from the wired interface of the test equipment to a wireless LAN interface is usually employed. Bridging may be accomplished with a computer having both wired Ethernet and wireless NICs (Network Interface Cards), and software to pass traffic back and forth between them. Very commonly, the bridge computer runs Linux; in this case, the standard Linux *iptables* routing software can be configured to perform the necessary bridging functions.

A simpler (though less capable) approach is to create a software tool capable of generating packet traffic with similar characteristics as the intended application, but not the same contents or transactions. An example of this method is the popular Chariot (now IxChariot) traffic generation software, which is capable of running on a wide variety of platforms, including laptops with WLAN cards. Built principally to allow IT staff to exercise their networks and diagnose performance issues, Chariot mimics the packet size distributions and data rates of common network applications (HTTP, FTP, VoIP, Oracle, etc.) without attempting to actually emulate the behavior of these applications. This approach is particularly useful when load-testing switching devices such as routers or gateways; it is not used for testing termination equipment such as servers or load-balancers, or stateful devices such as firewalls.

Whatever the test approach chosen, it is essential to have a stable, controllable and well-characterized traffic source in conjunction with an accurate and non-invasive traffic analyzer. Without these elements, application-level performance measurement becomes more of a hit-or-miss affair rather than a reliable test process.

6.1.7 End-to-End Testing

An end-to-end test is one where the whole system (traffic source, network, and traffic sink) is considered to be the system under test, and the results are reported for the entire combination. The following figure depicts this situation.

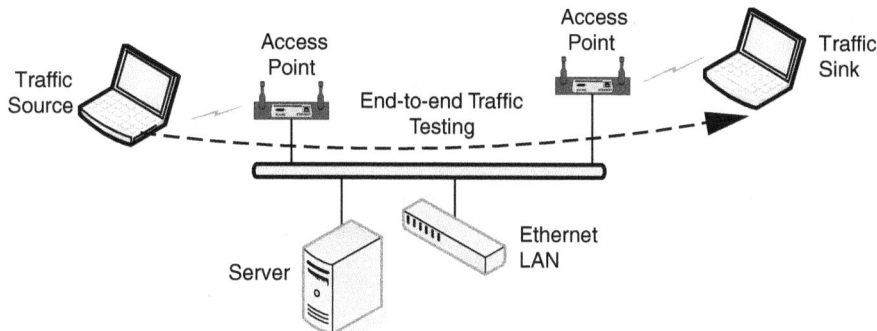

Figure 6.4: End-to-end Testing

In most cases the safest approach is to regard an application-level test as an end-to-end test. This avoids all issues with reproducibility or separation of test equipment issues from DUT/ SUT (system under test) issues. However, this is not necessarily palatable to the consumers of the test results, because an end-to-end test is only valid for that particular combination of equipment, and no other. This is perfectly acceptable when testing an installed network (including the laptops or computers connected to that network as clients) but not very useful when attempting to characterize a portion of the network, such as a single AP or a WLAN switch.

6.1.8 Link Segment Testing

A link segment test is basically directed towards an intermediate link (or hop) along the end-to-end path. It is conducted in much the same way as an application-level end-to-end test, but the test results are measured in such a way that the impact of the link segment of interest predominates. Figure 6.5 illustrates link segment testing.

An example of a link segment test is one that is conducted on an AP, using wireless source traffic and a wired Ethernet sink. In this case, the wired Ethernet interface of the AP can be safely regarded as not posing a bottleneck (the bandwidth on the wired side is at least 3X that of the wireless side) and hence what is being measured is the wireless performance of the AP. When conducting a link segment test with application-level traffic, however, care must be taken to ensure that the link segment is indeed the bottleneck, and not some other portion

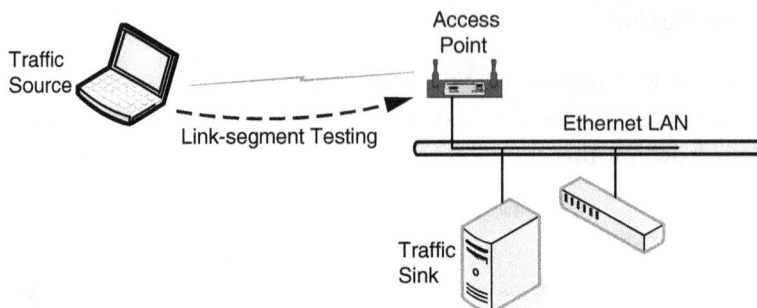

Figure 6.5: Link Segment Testing

of the end-to-end path (such as the traffic source). Otherwise, what is being measured is the capability of the test setup, not the capability of the DUT or SUT.

6.2 Application Traffic Mixes

Deployed networks actually carry a heterogeneous mix of traffic, with traffic streams of different characteristics flowing through the infrastructure, and in fact frequently being multiplexed down the same physical link. The insulation between the various classes of traffic is never as good as people would like to have them; it is not always possible to prevent voice packets from being delayed or dropped when bursts of less-critical data hit the WLAN switch or AP carrying them.

To avoid such problems, LAN equipment vendors resort to implementing elaborate Quality of Service (QoS) schemes to recognize and prioritize delay- and loss-sensitive traffic, and keep them from being affected by congestion. Testing with mixes of application-level traffic is therefore of significant interest, because they expose issues in these QoS implementations. The end-to-end nature of most application-level measurements lends itself particularly well to these scenarios, as QoS is a "whole-network" attribute rather than a link-by-link function. It is possible, by using traffic mixes, to get a complete picture of the characteristics of a LAN, whether enterprise or consumer; the picture includes the interactions of the various kinds of traffic at all points along the end-to-end path.

One caveat should be noted: rigorous benchmarking or equipment qualification procedures hardly ever use arbitrary mixes of application-level traffic. Mixed traffic flows usually result in even worse issues with repeatability than simple application-layer traffic, because the different flows evoke complex and frequently random behavior from the DUT or SUT. Further, the results obtained are not quite as clear-cut as tests performed with much more uniform traffic. For example, one DUT might perform well with one mix of traffic and poorly with another, while a different one could behave in the opposite manner; which DUT is better? This is obviously subjective and difficult to assess.

6.2.1 Mixed Voice and Data

Enterprise voice networks are transitioning to the use of packet voice (VoIP) instead of the analog or digital PBX setups now in widespread use. Many advantages result, ranging from a lower-cost unified network for all corporate locations to browser-based employee directories. However, the result of this trend is that shared-medium WLAN infrastructures end up sharing bandwidth between voice and data. It is therefore essential to determine the effect of mixing voice and data traffic over a single WLAN network.

Voice and data traffic streams have diametrically opposed requirements. Data streams cannot tolerate losses (e.g., losing a single byte from the middle of a file that is being transferred results in a corrupted file) and hence, are protected by protocols such as TCP that go to great lengths to maintain lossless data throughput. On the other hand, data traffic is quite insensitive to delay, and can tolerate wide swings in available bandwidth. Voice, however, is moderately loss-tolerant – small amounts of packet loss result in a drop in speech quality, not complete catastrophe – but needs a fairly constant allocation of bandwidth and is quite delay sensitive, both in terms of absolute delay and also jitter (delay variation).

Dealing with these conflicting requirements is therefore a challenge to equipment designers, as the datapath structures that produce the most efficient data transfers do not work well for voice, and vice versa. A considerable number of equipment compromises, shortcuts and assumptions are involved which make testing with mixed voice/data traffic quite important.

A typical voice/data test traffic mix may have from 5 to 10 voice calls and a few megabits/second of data per AP (most Voice over IP over WLAN (VoWLAN) vendors advise against overloading APs with voice calls, as voice quality drops off quite rapidly as the channel capacity limit is reached). Note that due to battery life constraints most VoWLAN handsets currently use 802.11b radios rather than 802.11g or 802.11a, so no more than 12–15 voice calls can be supported by a single WLAN channel, depending on the bandwidth requirements of the codec being used in the VoIP handsets. The following table gives the bandwidth requirements of some of the popular types of voice codecs.

Codec Type	Algorithm	Nominal BW	RTP Size	Packet Interval	RTP BW
G.711	PCM	64 kb/s	172	20 ms	68.8 kb/s
G.726	ADPCM	32 kb/s	92	20 ms	36.8 kb/s
G.723.1a	ACELP	6.3 kb/s	36	30 ms	9.6 kb/s
G.729	CS-ACELP	8 kb/s	42	30 ms	11.2 kb/s
Spectralink (proprietary)	PCM	64 kb/s	254	30 ms	67 kb/s

Note that the small VoIP packet sizes means that the actual proportion of channel capacity consumed by a single voice call is substantially more than what is represented by the above table, because of the relatively high overhead of 802.11 frames. For instance, a 172-byte Real Time Protocol (RTP) packet with ITU-T G.711 data (corresponding to a 228-byte

802.11 MAC frame) would require only 166 μs to transmit at an 11 Mb/s PHY bit rate, but the remainder of the 802.11 overhead (preamble, backoff, acknowledgement, etc.) requires 572 μs. Thus the data content of a VoIP stream is under 25% of the actual channel capacity consumed. In fact, the problem gets worse with more advanced codecs such as G.729 having a high compression ratio, because the packet size becomes still smaller.

6.2.2 Voice, Video and Data

Interest in transferring video – or, even better, a mix of video, voice and data – over 802.11 networks is spurred by the growing trend towards wireless multimedia links, primarily in consumer entertainment applications. There are also specialized vertical applications in the enterprise LAN space that are enabled by multimedia wireless links. Applications requiring digital video transmission include: videoconferencing, corporate training and video broadcasts, home entertainment, multiplayer games, video-on-demand and even airplane entertainment.

Video is not a significant portion of enterprise traffic at present, but the consumer market is very interested in video delivery – specifically, multimedia, which usually means a mixture of streaming video, audio, and data. Video traffic can tolerate small amounts of loss and jitter; absolute delay is not significant, but delay variation drives up the amount of buffering that must be maintained at the destination. Needless to say, successfully combining video, voice, and data requires that a lot of attention be paid to dealing with QoS and bandwidth efficiency issues, especially as the theoretical bandwidth of an 802.11a or 802.11g channel is no more than about 33 Mb/s – barely enough for a compressed video stream and some voice calls.

Video traffic demands a high and relatively constant allocation of bandwidth, ranging from 1.5 Mb/s for a low-quality MPEG stream to as much as 30 Mb/s for high-definition video.[1] This is in contrast to the approximately 70 kb/s each way for a single uncompressed voice link. The following table provides the typical bandwidth requirements of some commonly used digital video formats. Most compressed video formats use MPEG-2 today for performing the video compression and encoding, though the more advanced MPEG-4 standard is also coming into use.

Format	Typical Bandwidth	Remarks
MPEG-2	3–10 Mb/s	Can be as low as 100 kb/s and as high as 300 Mb/s
DVD	10 Mb/s	Most commonly uses MPEG-2 compression
HD-DVD	36 Mb/s	Primarily a storage rather than transfer standard
SDTV	15 Mb/s	Based on MPEG-2, 720 × 480 pixels @ 30 fps
HDTV	28 Mb/s	Wide variety of bandwidths; 12, 15, 18, and 39 Mb/s also used
DVB-C	6.4–64 Mb/s	Cable TV version of DVB; also based on MPEG-2
ATSC	19–39 Mb/s	Higher bandwidths usually found in CATV systems

[1] Transmitting uncompressed video over LANs is unheard of, as even a single Standard Definition Television (SDTV) would then consume >250 Mb/s.

The other aspect of video traffic is that the loss characteristics of the end-to-end link are quite important. Compressed video streams such as MPEG-2 and MPEG-4 are actually highly complex, with a repeating structure that covers many video frames (nominally 16 for MPEG-2). The repeating structure is referred to as a Group Of Pictures (GOP), Group Of Frames (GOF) or Group Of Video Object Planes (GOV). A GOF (or GOP, or GOV) is started by an intraframe (I-frame), which is the basis for all other frames within the GOF; subsequent to the intraframe comes a number of forward predicted frames (P-frames) or bidirectionally interpolated frames (B-frames), which are derived from the I-frame by means of a highly sophisticated compression algorithm. There are typically 16 frames in a GOF, of which the first frame is an I-frame and the remaining are P-frames and B-frames.

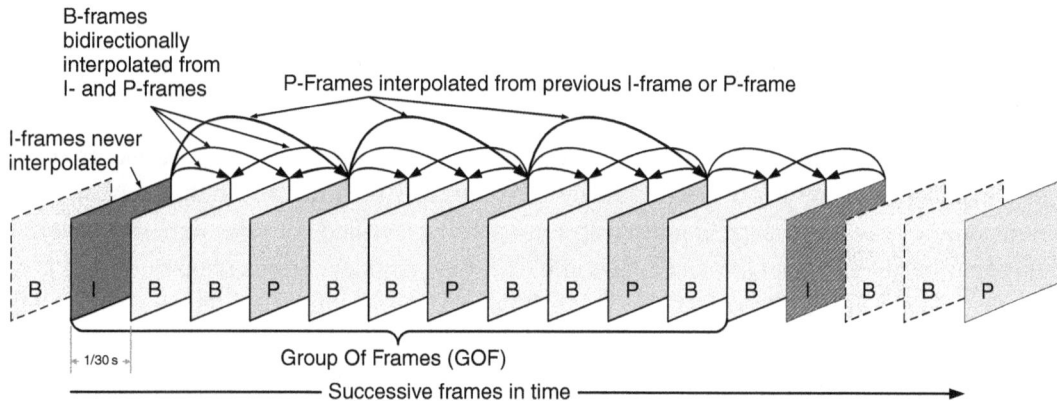

Figure 6.6: Structure of MPEG Video Stream

As the I-frame serves as the basis for the entire GOF, the loss of an I-frame causes significant and easily perceived defects in the video display after decompression, because the entire group will be lost. The loss of a P-frame is less significant than an I-frame, but still leads to visible artifacts because a block of subsequently transmitted B-frames will also be affected. Determining how well the network infrastructure defends video streams against losses is therefore an important test. Such losses are usually caused as a consequence of sudden bursts of data, or the starting of a number of voice calls; thus evaluating the QoS performance of the APs and switches is a key factor in predicting how the wireless network is going to handle some desired video load.

6.2.3 Control vs. Data Traffic

An often overlooked case that should be covered in such mixed-traffic scenarios is the impact of data traffic on control or management information traveling through the same links or channels. This is mostly of interest in large enterprise networks. For example, an enterprise wireless network transports not only user application data, but also network neighborhood discovery packets (e.g., to support printing and file service functions), beacons and radio resource management packets, Domain Name Service (DNS) queries, Universal Plug and Play (UPnP) information, and so on. The wired infrastructure backing the wireless APs further carries inter-AP and AP/controller traffic that allows the APs and controllers to exchange management and configuration information.

While the bandwidth required by this traffic is fairly low, it is critical to ensure that the traffic is allowed to flow unhindered, regardless of what may be going on in terms of data (or voice, or video) traffic. Otherwise, service interruptions are almost certain to result. For example, APs and controllers exchange "hello" packets to keep track of the status of the network elements and the topology of the infrastructure; if a sudden sustained burst of data causes these "hello" packets to be lost, then the APs are likely to disconnect from the controller and begin searching for a new one, causing the data streams passing through them to stop abruptly. In extreme cases, the whole infrastructure can cease to function while the individual devices try to re-establish connectivity.

It is therefore essential to verify that the devices and infrastructure can segregate and process control/management traffic quite independently of any realistic offered load that may be presented in terms of user data. Control traffic is usually classified with the highest priority, so this amounts to a test of QoS and bandwidth reservation under various scenarios (e.g., a common test is to bombard the DUT or SUT with application data traffic and then cause some disturbance, such as an AP being forced to reset, necessitating the exchange of control traffic; the response time of the SUT under these conditions is measured).

6.3 VoIP Testing

VoIP has been described as a "killer application" (to employ a grossly overused term) for wireless LANs in the enterprise. Certainly, there is a high level of interest from many enterprises in replacing the separate data and voice networks that exist today with a unified communications backbone, thereby realizing not merely cost and efficiency improvements but also new applications. However, as previously mentioned, successful deployment of voice networks over WLANs is highly dependent on QoS and bandwidth efficiency; thus adequate testing is key, primarily of the installed infrastructure but sometimes also of the devices and systems.

6.3.1 *Voice over IP over WLAN (VoWLAN)*

A typical corporate wireless VoIP system, as shown in the following figure, comprises (in addition to the normal wired LAN infrastructure) handsets, APs, WLAN controllers, a call server, and a PBX gateway. The PBX gateway is connected to the PBX, of course; a management console is usually present to allow the whole system to be managed from a central point. The presence of the WLAN component in the system has caused the entire ensemble to be referred to as VoIP over WLAN, or VoWLAN.

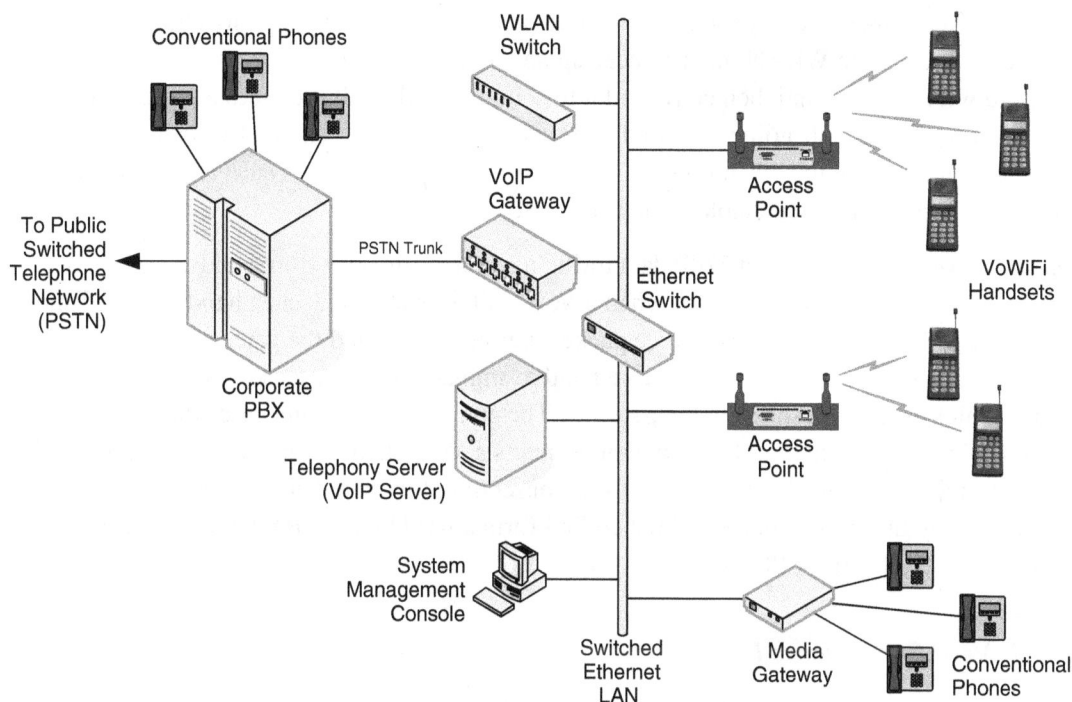

Figure 6.7: VoIP Over WLAN Architecture

In some situations the gateway, call server and PBX are rolled into one – the so-called "IP PBX" – but in most cases they continue to remain as two or three separate boxes.

In the above figure, the VoWLAN handsets function much like standard POTS (Plain Old Telephone Service) telephones: they convert between acoustic and electrical signals, provide a user interface, and connect to the network. A primary difference is that they encode and packetize speech, connect to an AP, and transmit the packets over an 802.11 channel to the call server or gateway. The call server implements signaling functions, allowing handset users to place and receive calls, using protocols such as H.323, Session Initiation Protocol (SIP) or MEGACO. Finally, virtually all enterprises have some pre-existing PBX system in place, so the PBX gateway converts between VoIP data streams and normal ADPCM (Adaptive Digital Pulse Code Modulation) digitized voice, and also allows VoWLAN users to access the public telephone system.

With the exception of some proprietary holdouts (e.g., SpectraLink), VoIP traffic today is carried using Real Time Protocol (RTP) over UDP/IP. Once the connection is established via SIP or some other means, the handsets digitize voice, encode using either a non-compressing codec (e.g., G.711) or a compressing codec (e.g., G.729), group sets of samples into RTP packets, encapsulate these packets in UDP/IP, and transmit them via the WLAN medium using

802.11 MAC frames. Exactly the reverse process takes place at the receiving end: packets are received from the WLAN media, decapsulated, converted into sets of samples, buffered, decoded with a codec, and then converted into sound. A VoIP call produces a more-or-less steady stream of packets going in both directions; silence-suppressing codecs can result in no data being transmitted when nobody is talking, but standard codecs such as G.711 create extremely regular and predictable packet streams.

One point worth noting is that VoIP data traffic never goes directly from handset to handset – i.e., bypassing the gateway. Instead, when a VoIP call is established, each handset establishes a separate bidirectional stream of voice packets between itself and the gateway; thus, if a VoIP traffic stream at a handset is captured and examined, all of the packets will appear to be coming from or destined to the gateway. The gateway, or in many cases the PBX, is responsible for the switching function that redirects a VoIP stream from one handset to another handset. If a PBX is involved, there is also a transcoding function that converts between the codec used by the VoIP handset and the coding format used by the PBX (standard A-law or μ-law ADPCM, for most PBXs).

6.3.2 VoIP Performance Factors

As VoIP service is principally for the purpose of allowing human beings to talk to each other at will, "performance" is admittedly a somewhat subjective term when applied to VoIP. Further, an extremely well-known yardstick of performance is available for user comparisons: the Public Switched Telephone Network (PSTN), which in most countries functions at very high levels of quality, reliability, and availability. Nevertheless, some measurable factors have been established as contributing to the perceived performance of VoWLAN systems.

The first and most important factor is, of course, voice quality; poor-quality voice calls negate all other advantages that a system may have. Quantifying voice quality is a complicated and subjective issue, but fortunately much research work has been done in this area and objective methods of scoring the quality of a VoIP call exist. VoWLAN systems have the further complication that call quality must be maintained while handsets are moving from location to location.

A second key factor is reliability, in terms of the dependability and uptime of the VoIP system. Enterprises have come to rely on their telephone service; in many cases, if the phone system goes down the corporation goes down with it. Thus quantifying the reliability aspects of the VoIP system, such as recovery time after a major outage, is of considerable interest.

Other, less tangible factors also exist. For instance, a great advantage of going to a unified voice/data network is the database and directory integration that can result; an employee's e-mail, phone, storage, and computing privileges can all be managed as a single unit. Of course, this has its cost in terms of load on directory servers and databases, with the result that factors such as call setup time, call capacity, call blocking probability, and so on are of interest.

6.3.3 VoWLAN-related Metrics

The following table summarizes the principal metrics that are used to characterize the performance of LAN infrastructure when supporting VoWLAN.

Metric	Description	Remarks
Call quality	Voice quality, expressed using some objective measure such as R-value (R-factor) or perceptual speech quality metrics (PSQM)	Typically related to delay, loss and jitter properties of the network infrastructure
Call capacity	Number of VoIP calls that can be supported before call quality drops below an objective threshold, or calls are refused	Controlled by the aggregate bandwidth of the network, plus bandwidth efficiency
Service level assurance	QoS maintenance by the network, expressed as the amount of lower priority data traffic that can be offered before call quality drops below an objective threshold	Controlled by the efficacy of the QoS mechanisms in the network devices, plus the queuing and datapath implementation
Service restoration	Time required to restore voice calls to an objective call quality level after a recoverable network outage, such as the loss of a primary controller	Usually related to the provisions for hot-standby and backup devices to guard against equipment failure
Roaming delay	Time required to restore an end-to-end VoIP connection after a handset switches from one AP to another	Dependent on fast roaming and other mobility enhancements; typically desired to be under 50 ms
Call setup rate	The maximum rate at which VoIP calls can be connected, which in turn measures call setup time	This is a control-plane rather than data-plane metric, and measures the performance of the call server and signaling stack
AP coverage	The radius of coverage (expressed either directly in distance, or in terms of path loss) for which an AP can maintain a specified call quality	Controlled by the efficacy of the RF layer, as well as the ability of the system to reject interference and recover quickly from bit error

Other metrics such as call completion probability or call blocking probability are also used when characterizing a VoWLAN system, though not as important as those previously described. However, the above table gives those quantitative measurements that are most directly related to the user experience.

6.3.4 VoWLAN Test Setups

VoWLAN test setups may be broadly divided into two categories: those that use "real phones" (i.e., actual handsets), and those that attempt to simulate the handsets using traffic generators. Both have their advantages and disadvantages.

The figure below shows a typical VoWLAN test setup employing real handsets as sources and sinks of traffic. In such a scenario, the entire infrastructure to be tested is set up (and acts as a SUT, as denoted by the dotted lines), and then some number of handsets are used to establish VoIP calls between themselves. Once the calls are established, various metrics such as call quality or service assurance can be measured.

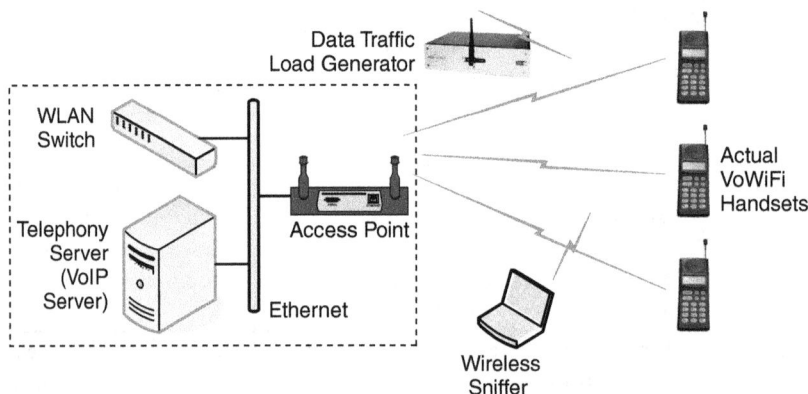

Figure 6.8: VoIP Test Setup – Real Phones

Passive sniffers, which are typically laptops with WLAN cards running packet capture software, are used at both ends of each VoWLAN link to capture all of the VoIP traffic being exchanged between the handsets. A post-processing step then extracts information from the capture files and outputs the desired metrics. A data load generator – a hardware or software means of injecting controlled amounts of data traffic – is used to set up a background load representing typical user data traffic for QoS related measurements. The handsets and APs may be placed in the open-air, or else RF enclosures may be used for a conducted test setup.

An alternative to real handsets is to use a hardware or software traffic generator and analyzer device to both source and sink traffic from emulated VoIP handsets, and to analyze (usually in real-time) the traffic in order to extract and present the desired metrics. The traffic generator may implement a more-or-less complete simulation of some number of handsets, depending on the test to be performed; in the case of call quality or QoS tests, it is only necessary to simulate the RTP streams carrying voice packets, while for call connection tests it is also necessary to simulate the SIP or H.323 stacks that are supported by the handsets. The test setup for this approach is shown below.

Telephony
Server
(VoIP Server)

Ethernet
Switch

WLAN
Switch

DUTs in Isolation
Chambers

AP   AP   AP

Many of VoIP
Phones Emulated
Per AP

Traffic Generator
and Analyzer (TGA)

Host
Computer
with Test
Software

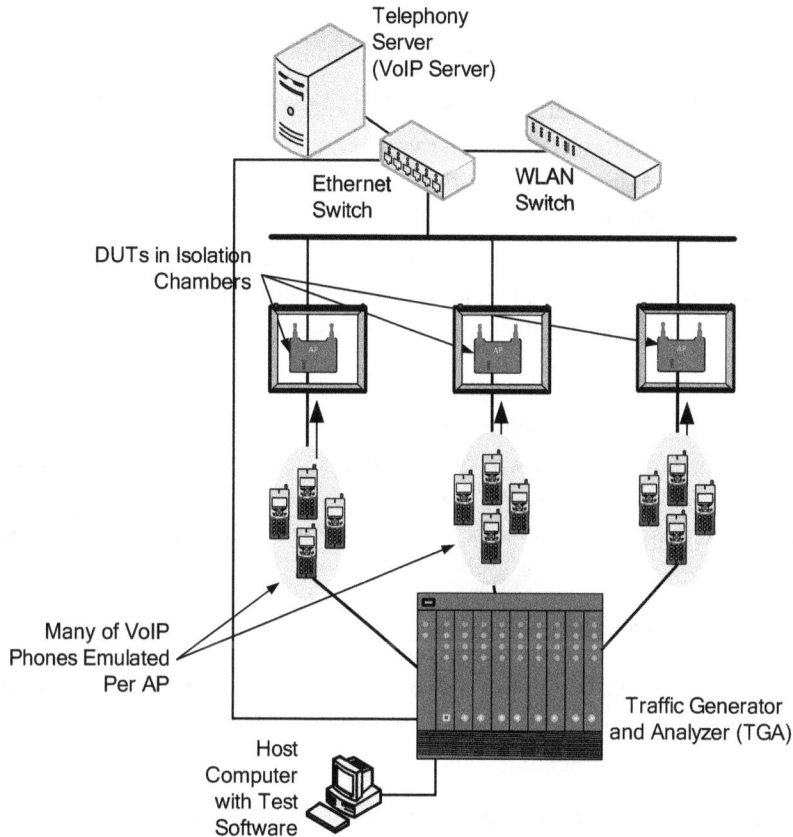

Figure 6.9: VoIP Test Setup – Emulated Traffic

The two approaches have their respective merits and issues. The first approach (real handsets) has the obvious advantage that the subtle interactions of actual handsets with the network infrastructure and each other will be captured in their entirety; such interactions have been demonstrated in actual tests to make a difference in perceived and measured performance. Further, there is no issue as to the validity of the results; emulation of handset traffic, however, almost always raises questions about how accurate the emulation actually is.

Unfortunately the use of real handsets also incurs a great deal of manual labor, particularly when setting up the test; for example, each pair of handsets must be connected with a manually-dialed call at the start of each trial. Further, it becomes unmanageably expensive for even moderate numbers of handsets (more than about 20); the open-air approach requires a great deal of floor space to prevent all the handsets interfering with each other, while a chambered setup necessitates an impracticably large number of enclosures and a lot of sensitive RF plumbing. Finally, the need to capture all the data and perform post-processing not only drives up the test time, but constrains trial durations to be quite short (under 5 min).

The second approach (traffic generator) is far more powerful, in terms of the amount of VoIP traffic that can be presented to the SUT, and the number of handsets that can be represented. Obviously, replacing the individual manually-dialed handsets eliminates all the manual labor during testing; but, in addition, it is only necessary to drive each AP with a separate interface on the traffic generator, and hence the test setup can be scaled to an extremely large extent. A test setup capable of emulating 500 handsets while measuring all of the metrics previously mentioned is not at all unusual. Further, measurement and analysis can be made real-time and automated.

The sole issue with the latter approach is that the handsets are emulated, and thus it cannot be guaranteed that the system behavior will exactly match that with real phones. However, the benefit of scalability makes this the only realistic option when testing the capacity of entire WLAN controllers, which may support more than 150 APs and thousands of handsets.

6.3.5 VoWLAN Measurements

Either of the two setups described in the above section may be employed to measure any or all of the metrics specified previously, though the emulated traffic approach is certainly more convenient for large-scale VoIP testing.

Call quality is straightforward to measure – simply set up the desired number of real or emulated VoIP streams (calls), analyze the traffic, and calculate an objective voice quality figure, most commonly in terms of an R-value (or R-factor). The R-value is not directly measured; instead, the key parameters contributing to it – end-to-end delay, jitter, average packet loss and burst loss – are measured, and the R-value calculated from these parameters using the equations and defaults supplied in ITU-T G.107. Some knowledge of the impairments and other factors for the specific voice codec being used or emulated is also necessary; these factors are provided in ITU-T G.113.

The R-value ranges from 0 to 100; values below 50 indicating unacceptable voice quality, while values above 90 signify very good quality. However, the R-value is an objective metric. It can be used to estimate relative call quality (or compare one VoIP network with another) but does not directly provide an absolute indication of subjective user satisfaction. User satisfaction measures are more commonly provided by what is known as a Mean Opinion Score (MOS), which is obtained by selected persons – the so-called "golden ears" – listening to the actual reconstructed audio, judging its quality and then assigning it a score from 1 to 5. Extensive research has been done to correlate the R-value with the MOS score, so that objective voice quality measurements can be used to gauge actual user response. The following table shows the relationship between an R-value figure and the corresponding MOS.

R-Value (lower limit)	MOS (lower limit)	User Satisfaction
90	4.34	Very satisfied
80	4.03	Satisfied
70	3.60	Some users dissatisfied
60	3.10	Many users dissatisfied
50	2.58	Nearly all users dissatisfied

Once call quality can be measured, measuring the coverage range of the AP for a given call quality is simply a matter of introducing attenuation between the handsets and the APs – either directly, by moving the handsets further away, or indirectly, by attenuating the signals or emulating the effects of path loss – and increasing it until the call quality drops below a threshold. This provides the coverage range of the AP, expressed either in terms of an absolute distance, or as the effective path loss over which the system can function at the desired level (See section 4.5.1 as well).

Call capacity is likewise straightforward. The number of real or emulated VoIP streams presented to the SUT is simply increased progressively, with call quality being monitored on all streams, until the quality of at least one call falls below a threshold, or the system refuses to accept any more calls. In some cases, a fixed level of data traffic may be simultaneously presented in order to represent some anticipated amount of background data load. Service level assurance (i.e., QoS maintenance) is a variation on this approach; instead of increasing the number of calls and keeping the data load fixed, the number of calls is held constant (e.g., at 1 call) and the data load is progressively increased instead, until the quality of at least one call falls below a minimum threshold.

The remainder of the metrics are principally control-plane measurements. Roaming delay is measured by shifting the real or emulated handsets from AP to AP while all of the calls are in progress, and measuring the time delay between the cessation of VoIP traffic on one AP and the resumption of VoIP traffic on the next one. It is important to ensure that the "true" resumption of end-to-end VoIP traffic is detected; in many cases the handsets may begin to send data on the new AP long before anything is received from the call server.

Service restoration time is measured by setting up some desired number of VoIP calls through the system, and then causing an outage (e.g., bringing down a primary controller). If the VoIP streams are interrupted while the system switches over to the backup controller, then the duration of the interruption is the service restoration time. The call setup rate is usually measured only with emulated handsets, as it is very difficult to perform this test repeatability with real handsets and humans dialing calls. The test itself is simple – new calls are presented to the SUT at progressively increasing rates until the SUT fails to complete one or more calls, at which point the maximum call setup rate has been found.

6.4 Video and Multimedia

Video traffic is currently of far more interest in the home and consumer markets than in enterprises, though it is possible that this will change as applications such as teleconferencing and remote learning become popular. Streaming video transmission of any kind requires a much higher sustained bandwidth than voice or data applications; even highly compressed video consumes megabits of channel capacity. This is not too difficult with switched wired LANs, where the pipes are large (100 Mb/s to 10 Gb/s) and dedicated to users, but with shared-media wireless networks it is somewhat of a challenge. Thus qualification of both individual devices (e.g., APs and WLAN controllers) as well as the complete system becomes highly necessary in order to guarantee a good user experience.

6.4.1 WLANs and Video

The typical application for video in a WLAN is a home multimedia center, where most of the wired cables have been replaced by wireless links. This not only enables the existing consumer applications to be supported, but also enables new ones resulting from integrating the computer and the Internet with the home entertainment center, such as computer-mediated video-on-demand and management of home video libraries. An example of a wireless multimedia server, and the network that would be built around it, is given in Figure 6.10.

As the bandwidth requirements of video transmission are very high, a new protocol mode referred to as Direct Link Setup (DLS) has been introduced with the adoption of the IEEE 802.11e standard. DLS enables client stations to bypass the AP and exchange data directly with each other, after an initial DLS connection setup handshake. This halves the channel

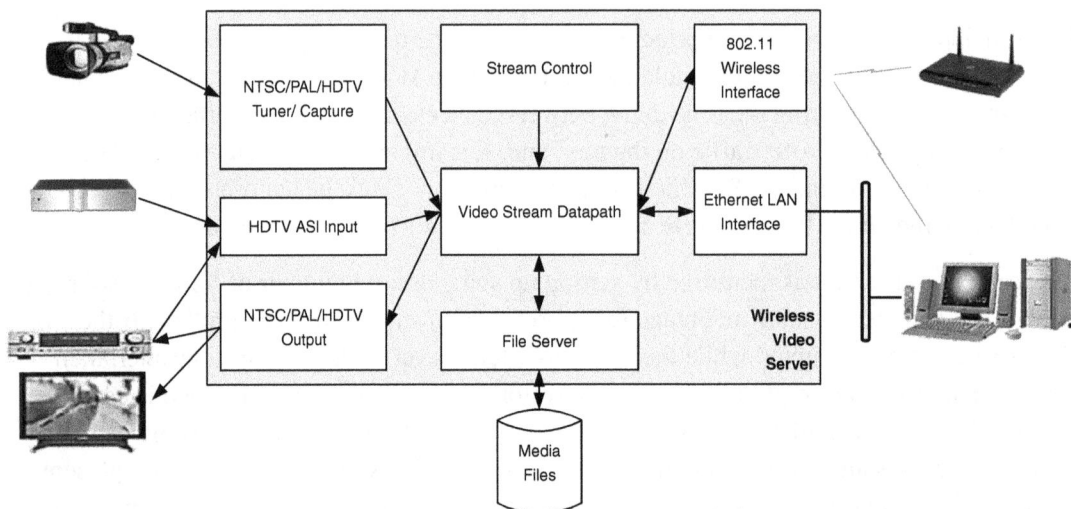

Figure 6.10: Wireless Video Server

bandwidth required for a video stream as compared to the situation where the video source sends packets to the AP and the AP then retransmits the same packets to the video sink.

Audio almost always accompanies video; in some cases, data (e.g., subtitles or menu information) is also transmitted concurrently. Hence the term multimedia. As the different media streams are sent over parallel but separate connections, it is essential that the end-to-end path preserve some degree of correlation between the timing of these parallel connections. Otherwise, the audio can arrive completely unsynchronized from the video.

With the exception of video teleconferencing, however, where a two-way real-time link is established, video applications can tolerate a moderate amount of latency. In fact the acceptable latency is substantially higher than that for VoIP. The latency limit for streaming video in the home is primarily set by disconcerting "time shifts" when moving between two TVs showing the same program, and is quite loose. In a commercial production and broadcast environment, however, the latency limits are much tighter.

6.4.2 Typical Requirements

The WLAN video user experience is affected principally by video quality, due to the bandwidth and interference limitations of WLANs. Unlike voice networks, objective metrics for video quality are still being developed (e.g., by the Video Quality Experts Group). The known factors affecting video quality are: sustained bandwidth, latency, jitter, and packet loss. All of these taken together affect the encoded and compressed video.

In addition to basic video quality, stream synchronization (between linked audio and video streams) is also important; failure to maintain stream synchronization to within milliseconds results in easily perceived "lip-synching" effects. The network capacity in terms of the number of simultaneous video streams that can be supported is also an interesting metric, especially in consumer situations where multiple parallel streams may have to be supported in the same room. Finally, the usable range is of interest, particularly as WLAN channel capacity drops quite precipitously once a certain level of path loss is reached, due to the rate adaptation functions implemented by the APs and clients.

The minimum requirements for a SDTV are summarized in the following table.

Parameter	Requirement	Remarks
Bandwidth	>14 Mb/s	Single SDTV stream
Latency	<250 ms	<100 ms for commercial applications
Jitter	<5 ms	
BER	$<10^{-6}$	After WLAN retransmissions are taken into account
Inter-stream synchronization	<120 ms	For commercial applications, <1 ms
Range	>20 m	Through the usual walls and partitions

Note that the above represents a minimum set; for example, a HT video stream would demand more than 28 Mb/s of sustained bandwidth, which reaches the theoretical maximum that a single 802.11a/g channel can provide.

6.4.3 Video Quality Metrics

Perceptual video quality is even more subjective than voice quality – the eye can detect extremely small defects or short-lived artifacts in a video image, whereas acoustic signal processing in the human auditory system is capable of compensating for all manner of noise when listening to speech. Considerable research is being done in the area of objective video quality metrics, but this field is still in its infancy.

One basic objective metric that has been proposed by Intel Corporation is the Gross Error Detector (GED). Basically, the GED expresses the quality of a received video stream in terms of the number of fairly serious errors (dropped frames and repeated frames) that may be caused due to network impairments. The GED is relatively easy to measure: an uncompressed source video stream (i.e., a selected video clip) is suitably instrumented by inserting markers (signatures) into each frame, then compressed and transmitted over the network. At the receiving end, the video stream is decompressed, and the markers are then extracted and analyzed. Analyzing the markers rather than the frames themselves makes it easy to identify gross errors – dropped or repeated frames. The total number of gross errors in the received video clip constitutes the GED.

To complete the process, the GED is correlated to subjective video quality scores assigned by human viewers, thereby providing a means to convert an objective quality metric into a subjective user satisfaction index. Note that this is the same process whereby an *R*-value for a VoIP stream is correlated to a MOS for audio quality, as previously described; in fact, the subjective video quality index is also referred to as a MOS, and also ranges from 1 to 5. This correlation is still under research, but some preliminary work indicates that the video quality MOS and the GED are related according to the following table.

GED	Video MOS
0	4.7
10	3.36
20	2.97
40	2.58
80	2.18
160	1.78

Another, more complex metric is the VQM, or Video Quality Metric, originated by the Institute for Telecommunication Sciences (ITS) and later taken up by ANSI (as ANSI T1.801.03-2003) and International Telecommunications Union (ITU) (ITU-T J.144R and ITU-R BT.1683). Rather than detecting gross errors such as dropped or repeated frames, this is based on actually analyzing the received video stream for human-perceptible artifacts such as blurring, "tiling" and noise. The analysis is then related to human (subjective) scoring of the displayed video to produce a correlation between the objective score generated by VQM and the subjective score of the human observers.

It is also possible to measure the individual components – available bandwidth, latency, jitter, loss, and burst loss – of the impairments imposed on a video packet stream by the WLAN infrastructure and thereby indirectly deduce video quality. If any of these components fail to meet the minimum requirements described in the previous section, then it is straightforward to conclude that video quality will be adversely affected. This type of indirect video quality assessment is encompassed, for example, by the Media Delivery Index (MDI) metric specified in RFC 4445. The only issue here is that expressing the human-perceived video quality degradation numerically, based on these measured impairment components, is not possible at present.

Inter-stream synchronization is a subject that is still under study, and hence no metrics are known to exist for this at the present time. It is quite difficult to measure objectively, as it depends on the correlation of two different media streams of completely different types. An indirect measurement may be made by comparing the packet latency and jitter experienced by the video stream with the same numbers obtained for the audio stream – a significant difference indicates the possibility of poor synchronization.

In addition to measuring video quality for a single stream, it may also be important to determine how many concurrent streams can be carried through the SUT before quality breaks down. As the typical 802.11a/g channel can carry only one or two video streams, this metric is more interesting for a SUT containing a WLAN controller and several APs, where multicast video is being transported to a number of clients. The measurement of the metric is simple, the process consisting merely of increasing the number of concurrent video streams progressively until the video quality metric for any of the streams falls below a threshold.

The range (coverage radius) of a WLAN AP while delivering video streams is quantified in exactly the same way as for voice streams. Once a metric for video quality has been selected and implemented, then the range is simply the path length (in terms of an absolute distance or an equivalent path loss) from the AP for which the video stream can be transferred with quality being maintained equal to or better than some preset threshold. In the case of a conducted test, the path loss is introduced by interposing a variable attenuator, or modifying transmit power.

6.4.4 Measuring Video Impairments

Video impairment measurement requires, in addition to the SUT, a video source as well as a video capture device (assuming post-processing) or a video analysis device (for real-time analysis). A display may also be part of the setup, for monitoring the received video stream.

The video generation and analysis may be performed by a traffic generator/analyzer, if it is capable of transmitting and analyzing compressed video clips with markers for the GED metric. The following figure illustrates the general setup.

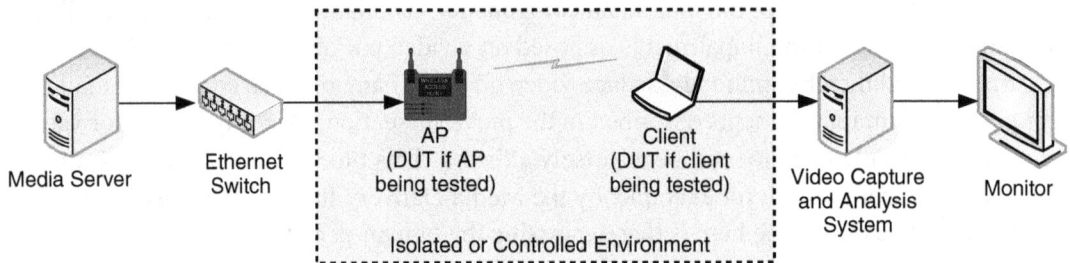

Figure 6.11: GED-based Video Quality Test Setup

Operation is quite simple. During the measurement of the desired metric, the video clip data is transmitted (as packets) from the source port (i.e., the video server, or emulation thereof); after traversing the SUT, it is received by the destination port (i.e., the client, or emulation thereof) and then either captured and stored, or analyzed in real-time. The video quality is obtained from the analysis and reported.

6.5 Relevance and Repeatability

The benefits of application-level testing should be fairly obvious from the above. The largest single issue with end-to-end application-level testing is as follows: unless the whole system model is thoroughly understood, and the test setup is created and characterized very carefully, the test results may have nothing to do with the SUT that is ostensibly being tested. Further, as previously mentioned, application-level test results may be irreproducible if the test equipment itself suffers from random and unrepeatable behavior.

Some of these issues arise from the complexity of the interactions between the DUT/SUT and the application-level traffic used in the test. However, with modern LAN equipment containing well-designed datapaths (frequently implemented in hardware) this is not a great problem. Much larger issues are caused, in fact, by the source and sink of the test traffic.

Consider for example the video quality metrics described above. Measuring video quality requires a video source (media server), which is frequently implemented as software running on a high-performance server or workstation, the video clip being stored on a hard disk. However, servers themselves are relatively unpredictable in terms of application-level delays (e.g., due to disk accesses, or even periodic virus checking) and can thus introduce impairments into the test traffic before it even reaches the SUT. Similarly, the video sink is a capture device, which is again usually a high-performance computer with a high-capacity storage medium such as a hard disk. Processing and access delays in the computer can easily cause frames to be dropped or delayed, irrespective of what happens in the SUT.[2]

It is therefore necessary to characterize the test setup extensively without the SUT in place – that is remove the SUT, connect the traffic source to the traffic sink, and run a number of long trials to verify that the behavior of the source and sink will not introduce measurement artifacts. If the source and sink do impact the measurement, then the level of artifacts should be characterized and factored into the uncertainty of the actual measurement. In the video quality measurement example used above, for instance, the video server should be connected directly to the video capture (or analysis) device and the test video clips transmitted, after which the quality metric should be derived. If the GED metric is being used, for example, then the number of errors found without the SUT in place must be added to the measurement uncertainty when the SUT is inserted.

[2] For this reason, companies such as Tektronix build specialized, hardware-based video traffic generators and analyzers. However, these are normally aimed at testing cable TV infrastructure, and not applicable to WLANs.

WLAN Manufacturing Test

This chapter deals with the test procedures and test setups that are typically used during the manufacturing of wireless LAN (WLAN) equipment. Manufacturing tests cover a range of aspects: radio calibration and alignment, screening tests during production to detect and weed out defective parts or systems, and post-production sampling tests for quality control (QC) processes. Tests performed during manufacturing and field failure analysis testing also fall into this category.

Note that chip (IC) manufacturing tests are not covered. Such tests are carried out with highly specialized equipment and usually performed using special features such as internal register scan logic, Built-In Self Test (BIST), and other functions specifically for chip testing; they have little to do with WLAN functional or performance test procedures. The focus in this chapter will be on board- or box-level manufacturing. The typical manufacturing flow (insofar as there is such a thing) will be reviewed, and then detailed descriptions will be provided of the equipment and processes used at each step.

7.1 The WLAN Manufacturing Flow

The details of WLAN manufacturing processes are both intricate and vendor-specific. While most system manufacturing processes follow the same general flow, different equipment manufacturers have widely varying processes, requirements, and setups. Also, the process is complicated by the fact that virtually all WLAN system vendors use third-party contract manufacturing services rather than building production lines in-house, and so the various steps in the process may take place at different locations. Rather than provide details of specific manufacturing processes, therefore, we will provide a general overview of a "typical" flow and then concentrate on manufacturing test aspects that pertain to WLAN devices.

The following figure summarizes a typical manufacturing process for WLAN equipment.

The steps in the usual manufacturing flow comprise:

a. *Parts sourcing*: The purchase, delivery, inventory, and sourcing of parts and subassemblies is a science all by itself. Sufficient stocks of such raw materials are necessary before volume production can begin.

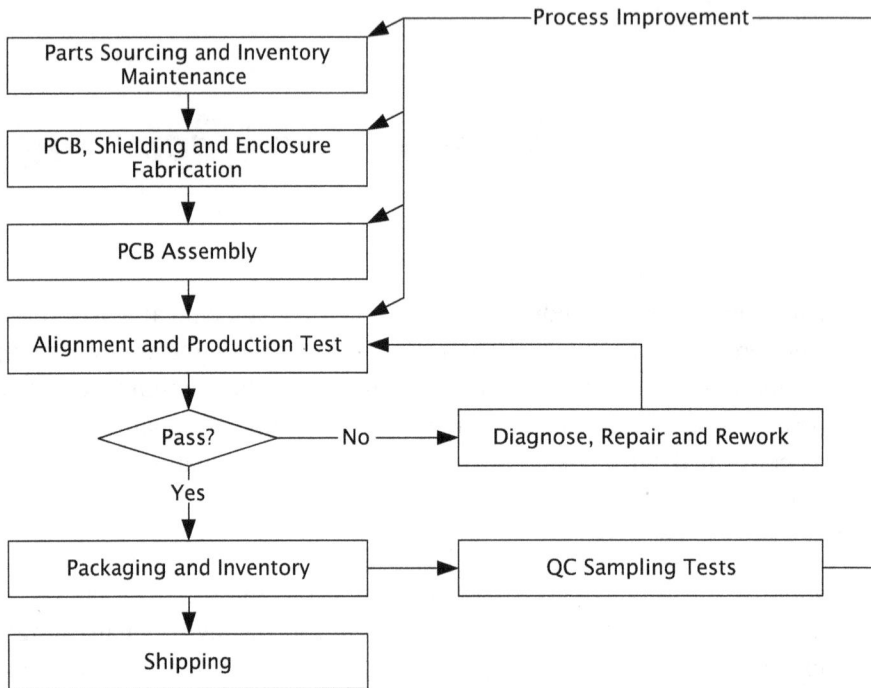

Figure 7.1: WLAN Manufacturing Process

b. *Creation of PCBs, shield cans, enclosures, etc.*: This may be done by the manufacturer, contracted to a supplier, or purchased off-the-shelf.

c. Assembly of parts on to Printed Circuit Boards (PCBs), and assembly of PCBs into subassemblies and finished products. Most of the machinery involved in high-volume system manufacturing (pick-and-place robots, solder reflow ovens, inspection stations, etc.) is associated with this step.

d. *Alignment, calibration, and production test*: This may be performed on the subassemblies as well as the finished products, and is the stage on which we will focus in this chapter. In the case of WLAN devices, it is common to align and calibrate radios within the subassemblies.

e. *Packaging and inventory*: Completed products that have passed the final production test phase are packaged in their shipping cartons and then placed on the shelf as finished goods inventory (FGI), before shipping to customers, distributors, and sometimes the equipment vendors themselves.

f. All high-volume manufacturing operations produce a certain percentage of defective goods, due to a variety of causes: defective components received from suppliers, short circuits or open circuits caused during the soldering process, failures introduced while handling or assembly, etc. (The ratio between total manufactured products and defect-free products is the manufacturing "yield".) When the cost of components and subassemblies warrants it, failures can be diagnosed and repaired ("reworked") after which the repaired products are subjected to the same production screening. If they pass, they are added to the FGI and subsequently shipped.

g. *Sampling tests for QC*: All high-volume manufacturers pull a certain percentage of finished and packaged products off the inventory shelves at random and perform more extensive testing. Such random sampling helps to identify manufacturing issues and deficiencies in the production testing that lets through defective products, and is a key part of ongoing quality improvement programs.

h. *Yield and process improvement*: Modern manufacturing houses continuously review and inspect all stages in their process, from sourcing to shipping, and seek efficiency (and therefore cost) improvements as well as yield increases.

Note that depending on the vendor, the production volume and the selling price of the product, there may be a stage of "burn-in" after production test and before packaging. This consists of actually running the product for some period of time (40 hours or more is not uncommon) and exercising its functions, and ensuring that the product continues to work without problems. Burn-in is done to weed out "infant mortality", as described later. Consumer-grade products selling in large numbers for low margins are usually not subjected to burn-in (i.e., dealing with infant mortality is left to the consumers!).

In addition to stocking quantities of raw materials (components, printed circuit boards, shield cans, etc.), manufacturers frequently stock finished subassemblies ("work in progress" or WIP) at different stages in the production line, to reduce delivery lead times without incurring extra cost by stocking finished goods.

7.1.1 Manufacturing Test Objectives

The efficiency and quality of a manufacturing process is quantified by the "yield" – as explained above, the ratio of good devices to the total number manufactured, after the final production test phase has culled out all the failures. Another useful metric is the field failure ratio, which is the ratio of devices that fail during actual use by the customer to the number of devices shipped; a high level of field failures can be disastrous for a vendor's profit margins and business. (Fortunately, field failures of WLAN equipment are rarely life threatening.) It should be noted that only failures within the design lifetime of the device should be counted for these metrics. The end goal of every volume manufacturing operation is to maximize yield, minimize field failures, keep manufacturing costs as low as possible, and minimize production times. A faster production process implies that the manufacturer can maintain lower inventories of both finished goods and WIP, and also respond more quickly to demand.

All manufacturing operations result in a certain percentage of defective products; a good manufacturing process can reduce the percentage, but it is not possible to completely eliminate them. The chief causes of defective equipment are:

- component variations and defects;

- manufacturing process variations;

- assembly defects, such as cold solder joints, misalignment of components, etc.;

- mechanical, thermal, or electrical overstress during or after manufacturing causing component or PCB failures;

- operator errors (e.g., programming in the wrong Medium Access Control (MAC) addresses);

- design problems resulting in manufactured products failing to meet specifications; and

- infant mortality due to early component failures.

As manufacturing defects can never be completely banished, screening processes are used to weed out defective products during production. These screening processes do not try to determine if the design is bad, or if the product does not have adequate performance; they instead look for failures caused by the factors listed above. Thus the tests used during manufacturing have a quite different objective from the tests used during all other phases of the WLAN lifecycle.

The primary objective of manufacturing tests is to identify and eliminate defective parts before they are shipped to customers. This is accomplished through different kinds of test processes applied at different stages:

- component/subassembly screening (not always done), where individual components or partially assembled modules are subjected to tests prior to being used for further assembly;

- production testing, where completed units are tested to ensure that there are no manufacturing defects;

- post-production sampling tests, where samples of manufactured and packaged units are thoroughly tested to ensure that the manufacturing process is sound and no unsuspected defects are slipping past the preceding test processes.

Tests are done at different stages to reduce manufacturing costs and to quickly catch defective manufacturing equipment or operator errors. The overhead incurred due to defects increases progressively as the product passes through the production phases; the cost of discarding or reworking a finished product is higher than the cost of dealing with a defective subassembly, and the cost of a field failure (i.e., after the product has left the factory) is the highest of all.

In a high-volume manufacturing operation, many of the manufacturing steps are performed in parallel (i.e., in pipelined fashion). Thus while it may take days or weeks for a single product to move through the manufacturing flow, the parallel operations mean that a new finished product will pop out at the end of the manufacturing line every few minutes or even seconds. Manufacturing test is squarely in the path of this high-volume process, and hence test time is very expensive; the expense is due to the cost of the floor space on the production line that is needed to house the test equipment, the labor cost for the operators to run the equipment, and the cost of the test equipment itself. All efforts are therefore made to reduce it.

Manufacturers go to great lengths to design test setups and test software to maximize test coverage and minimize test time. The production screening thus usually performs a fairly cursory test, rather than the comprehensive testing that occurs in verification and quality assurance (QA) labs. Much more comprehensive post-production tests are, however, done for QC and process improvement purposes. During this stage, the manufacturer redirects a small percentage (typically 2–10%) of manufactured devices for additional extensive testing in a separate area, and may also pull packaged devices off the shelf for such testing (to catch issues during packaging process). These tests are done in the manufacturing facility, but not on the production floor. They resemble the extensive testing performed during design and development testing processes, as the limited scope permits more time and equipment to be devoted to this activity.

7.1.2 Assembly Test

Assembly test is also sometimes referred to as in-circuit test or electrical test. Its purpose is simply to locate bad or missing components, electrical shorts or open circuits (e.g., due to cold solder joints), defective PCBs, etc. Access to the individual elements is accomplished in various ways, such as a "bed-of-nails" fixture (a fixture containing a number of probe pins that contact metallic elements or probe pads on the PCB) or a Joint Test Action Group (JTAG) scan chain (a daisy chain of shift-registers built into digital devices that can be used to access device pins and PCB traces). Once access is obtained to the PCB traces and device pins, defects can be located by driving them with low-voltage signals and looking for the appropriate responses.

Assembly test techniques are not specific to WLANs; the same techniques are used industry-wide for manufacturing many different types of systems.

7.1.3 RF Calibration and Alignment

Unlike digital communications equipment, wireless devices require a process of calibration and alignment before the transmit and receive paths in them will begin to function properly. The calibration process sets the radio to the proper channel center frequencies by determining appropriate constants to be programmed into the built-in frequency synthesizer(s); aligns internal filter passbands, amplifier gains and comparator thresholds to obtain the desired RF characteristics; and calibrates the received signal strength indications RSSI and transmit power control loops.

Formerly these processes were carried out using manual trimming of analog components such as trimmer potentiometers and capacitors. However, the use of digital transmit/receive chains on modern WLAN radio boards enables alignment and calibration to be performed by writing values to device registers. These values are then used to set parameters for digital filters and PLLs, as well as to drive D/A converters that set analog thresholds.

7.1.4 Device Programming

Virtually all WLAN radios contain an on-board serial EEPROM that holds radio calibration and alignment parameters, as well as other key information such as MAC addresses (for client network interface cards (NIC)) and product-specific options. For example, the on-board EEPROM may be used to customize a radio for a particular market by constraining the frequency bands of operation. The final step of the calibration and alignment process is programming the EEPROM with the calibration parameters.

In addition, WLAN controllers, APs, and NICs usually contain embedded processors that implement the higher-layer MAC and security functions. These embedded processors require operating firmware to be programmed into an on-board flash EEPROM. To ensure that the manufactured device is loaded with the latest version of firmware, the programming of the flash EEPROM is done on the manufacturing line.

7.1.5 System-Level Testing

After the system has been calibrated and programmed, it can function as the final product. At this point, functional testing is carried out for a final manufacturing test, to ensure that the right passive components have been assembled and there are no partial failures in the active devices. (Assembly tests only verify that there are no opens, shorts, or completely non-functional parts; errors in component values, partially failed components, or programming errors are caught after the system has been fully assembled and system operation can be checked.) Functional tests are also a useful check on the quality of the overall manufacturing process, and the results are often recorded and analyzed to detect unwanted process variations.

Functional test is limited by manufacturing economics. Most functional tests are fairly superficial and limited to what can be performed with simple test setups in a very short time (under 30 seconds is usual). Sometimes a system test, i.e., a modified subset of the QA performance tests, may also be run to ensure that all the interfaces are working properly, and there are no hidden RF issues. If certain datasheet performance parameters are guaranteed during manufacturing, then these tests are run at this time as well, and the results recorded.

7.2 Manufacturing Test Setups

Manufacturing test setups have clear-cut objectives, as follows:

- They must be compact, in order to take up as little expensive floor space as possible on the production floor. This is particularly important in high-volume production, where multiple test stations may be used per line to increase production rates.

- They need to be labor efficient, so that they can minimize operator fatigue and maximize production rate.

- They are usually highly automated; this not only reduces test and alignment time, but also considerably reduces human error. Also, an automated test setup can maintain a centralized database of manufacturing parameters that is useful for process control.

- They should be easily reconfigured. Manufacturing lines have to be flexible, in order to keep up with product revisions and new product introductions.

- While cost is not the paramount consideration, they should be of moderately low cost in order to allow parallel setups to be employed to increase production rates.

A manufacturing test setup is quite unlike the equivalent laboratory setup, even though some of the same instruments may be used. The test setup is usually referred to as a manufacturing test station, and is almost always rack-mounted for easy reconfiguration and upgrade. In addition to test equipment, each test station contains a fixture for fast device under test (DUT, i.e., manufactured device) insertion and removal. Further, the test station is self-contained, so that all calibration/programming/test functions can be carried out in one step and at one location, and careful attention is paid to station construction to allow an operator to handle more than one manufacturing test station at a time. It is not uncommon to find one operator in charge of up to four adjacent manufacturing test stations, even though the total test time at any one station may be 60 s or less per device. Manufacturing test setups are frequently home-grown (custom-developed by in-house engineering belonging to the vendor of the manufactured device), but test equipment suppliers such as Agilent Technologies also provide "plug-in" integrated solutions that can be customized for specific manufacturing operations.

Figure 7.2 shows a typical setup for manufacturing test of WLAN client cards or small APs.

A manufacturing test station contains the following:

- A test fixture to support the DUT, and enable the DUT to be connected to the test station (and disconnected from it) very quickly.

- Microwave signal sources and signal analyzers for tuning and alignment, plus an RF power meter for calibration, and a DC power supply to power the DUT.

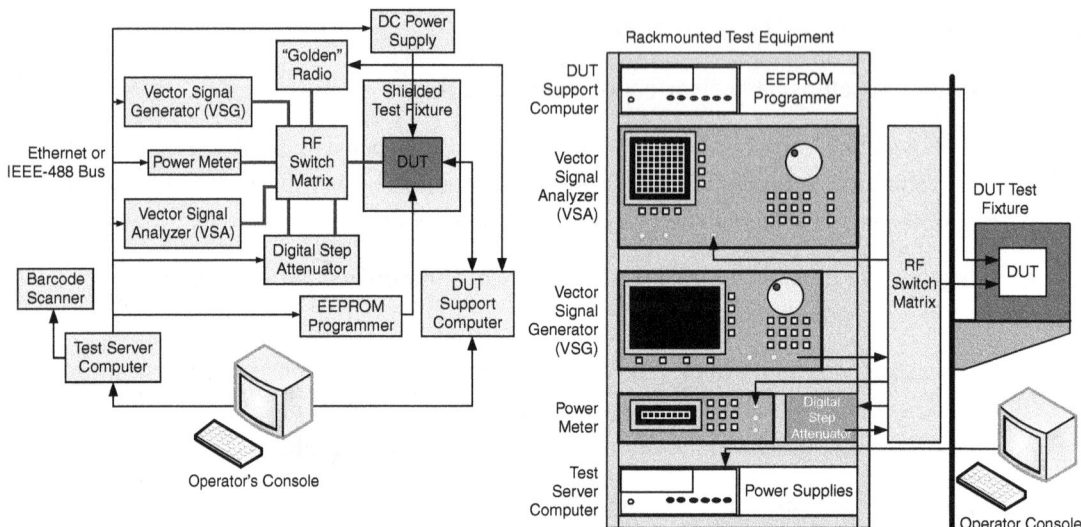

Figure 7.2: Manufacturing Station for Calibration, Programming, System Test

- A PC to control the DUT as well as to source or sink traffic during testing.

- An EEPROM programmer to configure radio parameters, write the MAC address(es) assigned to the DUT into the serial EEPROM, and also download DUT firmware into flash EEPROM(s) on the DUT PCB if necessary.

- A system control PC (also known as a "test server") to control the test equipment and communicate with a central database server to upload test records.

The interface between the test equipment and the system control PC is frequently via the General Purpose Interface Bus (GPIB, also known as Institute of Electrical and Electronic Engineers (IEEE) 488) or using RS-232C or RS-485 serial ports. Of late, though, Ethernet has been adopted as a universal test system interface, promoted by both test equipment vendors and manufacturers; this is standardized as LXI (LAN Extensions for Instrumentation), and allows bulky GPIB cables and unwieldy serial console multiplexers to be replaced by a simple and high-speed LAN switch. The interface between the DUT control PC, the system control PC, and the central manufacturing control and database server is via an Ethernet LAN, almost always running TCP/IP (though the author knows of at least one NetBEUI installation). New DUT firmware and new test software are downloaded from the central server, while test reports and calibration records are uploaded to the server as well.

The manufacturing test station involves quite a bit of software, both for presenting a simple user interface to semi-skilled operators as well as for automating the complete test process. The software for every manufacturing line is different – in fact, different software loads are used when testing different manufactured products. It is normally developed by the WLAN device vendor's manufacturing engineers, in conjunction with their QA and production teams; occasionally the manufacturer or manufacturing contractor may provide assistance as well. The software is quite complex and does many disparate functions automatically: it controls the ATE (Automatic Test Equipment) interfaces on the test equipment, downloads image files to device programmers, controls host computers and servers via their network interfaces, conducts the calibration/programming/test process, and records the results.

The central database contains all of the calibration, programming, and performance test information pertaining to each and every manufactured device. It is extremely valuable for performing trend analysis, spotting developing defect patterns, carrying out quality improvement programs, and detecting operator errors (such as misprogrammed MAC addresses) early. The database also allows remote monitoring of the production process; the production line is frequently outsourced to a contract manufacturer in a different country or continent from the engineering operations of the actual vendor of the DUT, and remote monitoring is essential to allow the vendor to track the progress of build orders and monitor the production processes.

Note that assembly test is not performed on a manufacturing test station; instead, special in-circuit testers (ICTs) are used.

7.2.1 "Home-grown" Test Stations

Due to the specialized requirements of different vendors' products, or even different products produced by the same vendor, it is quite common for WLAN equipment vendors to design and build their own manufacturing test setups. These setups, however, typically use off-the-shelf test equipment for the most part, with only a few pieces being actually constructed "from scratch"; most of the design task is a process of integration and software developments. Home-grown manufacturing test stations are designed by the manufacturing departments of the equipment vendors, implemented and validated on prototype production lines, and then installed on the production floor, possibly at a remote contract manufacturing site.

A typical "home-grown" manufacturing test station contains the following:

- An RF vector signal generator (VSG).

- An RF vector signal analyzer (VSA).

- An RF microwave power meter for transmit power and RSSI calibration.

- A shielded DUT fixture to quickly mount and dismount the DUT, support it, connect to its connectors, and couple RF signals to its antennas or antenna connectors.

- A precision programmable power supply to power the DUT.

- A remote-reading DC voltmeter and ammeter to measure DUT DC power consumption (sometimes this function is built into the power supply).

- Remotely controllable RF switches, combiners, etc. to connect the DUT to different devices in different ways (i.e., "RF plumbing").

- A programmable attenuator for power and dynamic range measurements.

- An EEPROM programmer to program the serial flash EEPROMs on the DUT.

- For client card testing, a PC interfaced to the DUT fixture to host the client device driver and OS SW.

- For AP manufacturing, a traffic generator/analyzer to run traffic through the DUT.

- A barcode scanner to allow the operator to read the unique ID number associated with the DUT – for example, to associate a specific MAC address with the DUT, and also to track the DUT through the production process.

- A system control PC to control the test equipment and switches, contain the test software, etc.

The operation of the typical test station described above follows a fairly well-defined sequence. The DUT frequency synthesizer is tuned first, to ensure that the transmitter and receiver channelization is correct. After that, the transmit and receive chains are aligned, and the necessary calibration steps (for transmit power and RSSI) are performed. An error vector magnitude (EVM) check may be performed to verify that all is well before proceeding to the next step. After this, functional test is carried out to verify that the system and modules are working properly. System test is usually done as a subsequent stage of functional test. Any failures at the calibration, EVM, functional or system test stage cause the product to be routed back into the manufacturing line for rework.

Home-grown manufacturing test stations normally make use of a "golden radio" to cause the DUT to generate and receive signals, once the DUT radio is active and the test process needs to start traffic flowing through it. For a client, the "golden radio" is usually a specially selected AP; for an AP, it is normally a client card in a PC. In both cases the manufacturer is forced to acquire a selection of such devices and manually select the ones that are of acceptable quality; WLAN client cards and APs are not designed as test equipment and have wide variations in critical RF parameters. In fact, one of the issues with a home-grown setup is the tendency of the "golden radio" to be not quite so "golden", and instead exhibit artifacts and irregularities that in turn lead to false positives or false negatives during the manufacturing process. (Either outcome leads to issues with quality and manufacturing cost.) To ensure optimum performance on the production floor, the entire test setup must be calibrated frequently, sometimes once a day, in order to ensure that manufactured products have consistent quality. The need for calibration is exacerbated by the heavy use to which such setups are put; for example, a typical test station may process one device every 30 s, and run continuously over two 8-hour shifts, resulting in over 2000 test operations per day.

7.2.2 Off-the-Shelf Setups

As the WLAN market is growing, test equipment vendors have started to bring out dedicated manufacturing test setups specifically designed for WLAN production test functions. These are essentially integrated versions of the home-grown test stations, containing many of the same capabilities, but sold as a unit rather than individual components. An example is the Agilent Technologies GS-8300 WLAN manufacturing test system. These test setups contain, in a single system:

- shielded test fixtures;

- all signal generation and analysis functions to enable calibration, alignment and tuning, replacing the laboratory-type VSAs, VSGs, power meters and RF plumbing;

- an integrated computer for system software support, calibration, and diagnostics.

As every manufacturer's test requirements are different, these dedicated manufacturing test setups are also accompanied by substantial amounts of customization and applications development support in order to adapt them to the specific needs of each production line.

7.3 Radio Calibration

Newly manufactured digital devices either work or do not work; there are no adjustments or "tweaking" that can make them work better. Defective digital devices are sent directly into diagnosis and rework operations. Unlike digital devices, however, RF equipment works poorly (or not at all) until calibrated and aligned. In the case of WLAN radios, the calibration and alignment process essentially determines a set of compensation and threshold factors for the various tunable elements of the transmit and receive chains, and then loads these compensation factors into the device or submodule. Modern WLAN radio alignment is a completely digital process, which is outlined in the Figure 7.3.

The actual calibration process is highly dependent on the type and design of the radio, and is determined by the manufacturer of the chipset. Most chipset vendors provide calibration procedures and even software packages to enable system vendors to calibrate their radios. Calibration takes the following general steps:

1. The crystal-controlled synthesizer is configured to match the channel center frequencies, by calibrating the crystal oscillator and then determining the appropriate PLL division ratios.

2. The transmitter chain is aligned for I/Q balance and spectral mask filtering. In addition, compensation constants are measured for the transmit power control loops.

3. The receiver chain is aligned, including setting constants for automatic gain control (AGC) operation and filter passbands. The RSSI is calibrated and adjusted to match the

```
┌─────────────────────────────────────────────┐
│  Calibrate the frequency synthesizer for the  │
│        desired channel center frequencies      │
└─────────────────────────────────────────────┘
                      │
                      ▼
┌─────────────────────────────────────────────┐
│ Align the transmitter and determine the various│
│           gain and I/Q balance constants        │
└─────────────────────────────────────────────┘
                      │
                      ▼
┌─────────────────────────────────────────────┐
│  Align the receiver and determine the RSSI and  │
│                 AGC constants                    │
└─────────────────────────────────────────────┘
                      │
                      ▼
┌─────────────────────────────────────────────┐
│   Determine LNA and diversity switching constants│
└─────────────────────────────────────────────┘
                      │
                      ▼
┌─────────────────────────────────────────────┐
│  Calculate and program calibration constants into│
│                on-board EEPROM                   │
└─────────────────────────────────────────────┘
```

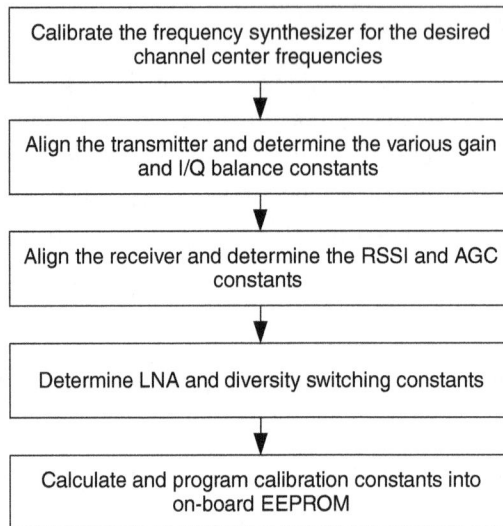

Figure 7.3: Calibration Process

datasheet specifications. Also, threshold values are measured for the LNA in/out switching and diversity switching levels.

Calibration is carried out completely electronically, by writing different values to registers within the chipset(s); in fact, in most cases a serial EEPROM is loaded with calibration values that are then automatically loaded into the chipset registers on power-up. There is no need to adjust potentiometers or capacitors, or any form of mechanical tuning or trimming. Internal D/A converters in the RF/IF chain are used to convert the register values into the actual RF parameter modifications; for instance, instead of altering a potentiometer to adjust gain, a variable-gain amplifier may be controlled by a D/A converter. The use of fully digital basebands further simplifies this because all of the filter tuning and I/Q calibration can be done digitally; compensation coefficients are loaded into registers and used during digital signal processing.

It is possible for WLAN devices to fail the calibration process (e.g., if no combination of parameters can be found that brings the RF performance of the transmitter or receiver into acceptable tolerances). Such devices are sent immediately to the rework stage.

7.4 Programming

During the programming phase, the on-board EEPROM(s) are programmed with the MAC address assigned to the interface (in the case of NIC devices), the product-specific chipset options, and the calibration constants determined during RF alignment and calibration. In addition, the usual practice is to use an embedded CPU within the WLAN chipset for the more complex upper-layer MAC and security functions; the chipset vendor may provide a firmware image for this embedded CPU that has to be programmed into a flash EEPROM. During manufacture, a bar-code strip is usually affixed to the module or PCB containing a unique serial number assigned to the module. (This serial number may sometimes be the MAC address to be programmed.) A bar-code scanner is used as part of the manufacturing test setup to read the bar-code and convert it into a globally unique MAC address, which is programmed along with the other information into the on-board EEPROM. The contents of the EEPROM are usually read by the host device driver and loaded into the device calibration registers every time the system boots; alternatively, the EEPROM may be automatically loaded by the chipset on reset or power-up. Until the registers are loaded, the chipset and therefore the system is non-functional.

7.5 Functional and System Testing

Functional and system tests are performed after the WLAN device has been fully aligned and programmed, that is, when the device is expected to be fully functional. These tests are conducted on a strictly pass/fail (go/no-go) basis, and are done to screen out defective parts from good parts. No test results beyond the binary pass/fail decision are reported or recorded (though in some cases the calibration information may be retained for process improvements). If the device fails functional or system testing, it is sent for diagnosis and rework; the rework process will do more extensive testing, but this time actually measuring and recording values in order to determine the probable cause of failure so that it can be fixed. Suitable thresholds must be built in so that an unduly large number of false positives (resulting in good parts being rejected) will not occur, while at the same time guarding against letting defective parts through. These thresholds are tuned over time as the manufacturing tolerances are tightened up (or loosened); manufacturers continuously update and improve their production processes, including test, to enhance manufacturing yields and lower costs.

Functional tests in the case of WLAN devices are mainly RF-level tests. The digital portions of the product are exercised by means of system-level tests.

7.5.1 RF-Level Functional Tests

A number of RF functional tests are carried out on finished WLAN products to ensure that they will provide good performance in the customer's hands, to guarantee compliance to regulations set by the Federal Communications Commission (FCC) or other governmental bodies, and to verify that the production calibration and component tolerances are acceptable. These functional tests are usually a small subset of the RF tests performed in the laboratory during design and verification. They include the following:

- Channel center frequency accuracy checks performed on every usable WLAN channel, to verify that the synthesizer has been adjusted and is working properly.

- Carrier suppression and EVM tests to ensure that the transmitter is generating signals of adequate quality.

- Spectral mask compliance checks to verify that the device meets standards and regulations.

- Transmit minimum/maximum power tests to ensure that the output power matches the datasheet specifications.

- Receiver sensitivity, maximum input level, and PER measurements to verify that the receiver parameters are within specifications.

- RSSI tracking tests, which confirm that the RSSI calibration is valid over the receiver input range.

- An antenna diversity check to ensure that diversity switching is working correctly.

7.5.2 System-Level Tests

System-level tests are performed on the manufactured product as a whole, rather than just the radio, to ensure that the remainder of the system or module is fully functional. Obviously the range of system-level tests can be quite broad and varied, and is also dependent on the nature of the system; for example, an AP will be subjected to different sets of tests than clients.

Typical system-level tests are as follows:

- Measurement of operating and standby DC power consumption; DC power consumption much above or below predetermined thresholds can serve as a quick and accurate indication of defective or malfunctioning devices.

- Verification, usually by reading identification registers on chips and PCBs, that the chipset and other revision numbers match those expected.

- Measurement of a few simple data-plane performance factors such as throughput or forwarding rate; this serves as a good indicator of overall health of the device.

7.5.3 QC Sampling Tests

In most manufacturing processes, particularly high-volume operations, a small sample (2–10%) of manufactured devices are diverted to more rigorous and detailed testing. These samples are taken directly from the end of the production line, or even from batches of finished and packaged goods awaiting shipment. Such QC sampling tests are done to ensure that the actual production floor tests have good coverage of faults and failures, and that defective parts are not slipping through. Also, they identify issues that may be occurring between the production test phase and the packaging phase (e.g., overstress or damage during the packaging process). Finally, they provide feedback to the manufacturing team, that can be used to tighten up production tests and improve the manufacturing process.

QC sampling tests are similar to the production testing, but are more elaborate and extensive. For example, the production test might simply measure transmit EVM at a specific rate, channel and power level, in order to save time. The QC sampling test, however, would measure EVM at all combinations of data rates, channels, and power levels; this takes a long time, but gives a much better picture of the performance of the transmit chain. If the production test indicates that the device passed, but the sampling test discovers a failure, then there is a manufacturing process problem that needs to be fixed quickly.

Due to their comprehensive and time-consuming nature, QC sampling tests are usually run manually on a laboratory-quality setup, away from the production test floor. For a production test floor with 100 manufacturing test stations, there may be 2–3 QC sampling test setups assigned. Equipment and test procedures used during these tests closely resemble those used in the manufacturer's design and verification laboratories.

7.6 Failure Patterns

As the primary purpose of manufacturing tests is to avoid having failed products in customers' hands, it is worth taking some time to understand failure patterns that affect all electronic equipment, including WLAN devices. The incidence of device failures is not uniform, that is, exhibiting a constant rate of failure over time – but instead forms a U-shaped pattern referred to as a "bathtub curve". The following figure illustrates the shape of the bathtub curve. The horizontal axis represents time, while the vertical axis represents the incidence or probability of failure of a particular piece of equipment.

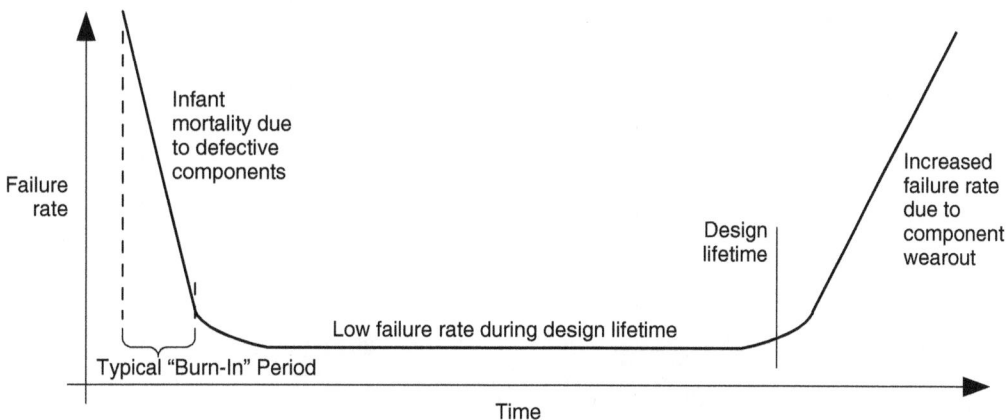

Figure 7.4: Bathtub Curve

As seen by the above curve, the initial failure rate is usually comparatively high; this is referred to as "infant mortality". It is caused by the fact that if there are defective or marginal parts in the system, they will usually fail catastrophically during the first few hours of operation. After the infant mortality period is over, the failure rate drops to a relatively low value; the products with marginal components have been weeded out by this time and the remainder will be operating normally, within their design parameters. As the design lifetime of the product is reached, however, the failure rate starts to rise again, because the components and submodules within the product eventually start wearing out and failing; this is referred to as "wearout failures" and is normal and expected.

Part of the manufacturing process is to eliminate infant mortality as a cause for field failures. If the manufacturing volume permits, it is common to use a period of "burn-in" (i.e., running the WLAN device with power applied and normal traffic passing through it) – to deliberately cause marginal devices to fail before they are shipped to end customers. If the volume is too high (and the selling price of the device too low), it is too expensive to spend the required amount of time on burn-in; in this case, careful up-front system design must be performed to build margin into the system, in order to cope with early field failures.

Another goal of the manufacturing operation is to ensure that the wearout failure rate only rises at the end of the design lifetime, thereby avoiding user dissatisfaction and excessive warranty costs. The primary means of ensuring an adequate design lifetime is to select components and materials with enough margin built in to ensure that the product can tolerate stress of use until the design lifetime is exceeded.

Installation Test

A considerable amount of testing is performed during the planning and installation of large- and medium-size enterprise wireless LAN (WLAN) installations. For example, site surveys are performed prior to installation, and performance qualification testing is done after the equipment has been installed, but before it is allowed to carry live traffic. A wide variety of techniques are in use, along with an equally wide range of low-cost test equipment, during WLAN installation, qualification, and maintenance.

This chapter describes the essential test procedures during the WLAN installation and maintenance process, as well as the types of tools employed. Note that it is not intended to be a primer for WLAN deployment (this is already well covered in many other books), but focuses on test equipment and methodologies.

8.1 Enterprise WLANs

This section briefly describes the architecture and components of an enterprise WLAN, and the factors to be considered before and after installation. A previous chapter (see Section 6.1.2) also provides some information about enterprise WLANs, and is worth consulting before beginning this one. Enterprise infrastructure equipment, as well as the general enterprise WLAN deployment process, will be summarized for clarity; however, only the testing aspects of deployment will be covered in detail.

8.1.1 Infrastructure

The initial deployments of WLAN equipment in the enterprise were relatively small, and used stand-alone access points (APs) connected into the wired LAN backbone; basically, these were merely "wireless extension cords" for the wired infrastructure. This is the distributed approach to WLAN implementation. As the adoption of WLANs increased and the average size of each deployment grew to hundreds rather than dozens of APs, a centralized approach proved to be more manageable and usable; in this situation, the APs are no longer stand-alone, but controlled by a central "WLAN switch". Today, all but the smallest enterprise WLANs are built with a centralized rather than a distributed WLAN architecture.

Centralized control greatly simplifies management of the wireless network as well as the security and Quality of Service (QoS) infrastructure required in an enterprise LAN.

It also simplifies some of the installation and maintenance problems that were encountered with large networks of stand-alone APs, by enabling the network to quickly adapt to changing conditions and requirements without forcing the corporate information services staff to access and reconfigure hundreds of APs. In some cases centralized control extends all the way from remote offices to corporate data centers; WLAN APs and controllers at branch offices may be configured and managed remotely, over a WAN link, from a corporate main office. Thus an IT staffer sitting at a console can monitor, diagnose, reconfigure, and repair the entire WLAN for a large corporation.

The following figure shows the infrastructure components typically used in a medium-sized enterprise WLAN system.

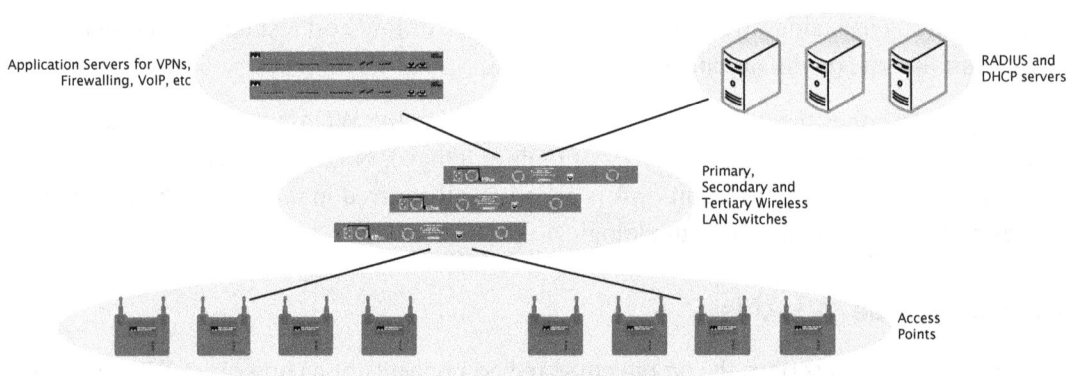

Figure 8.1: Enterprise WLAN Infrastructure Components

The above figure does not show the wired LAN infrastructure that connects all of the components; this is assumed to be in place and providing sufficient bandwidth and connectivity. (A typical Ethernet LAN installed by every mid-size or larger enterprise has more than enough capacity to support all the WLAN infrastructure that the enterprise may require.) Instead, the focus is on WLAN-specific components that are usually part of every installed WLAN, such as:

- APs;

- WLAN controllers (usually organized in a hierarchy of primary, secondary, and tertiary, as needed to handle the system load);

- an Remote Authentication Dial In User Service (RADIUS) server for security and authentication purposes;

- a DHCP server to support dynamic configuration of clients such as laptops and handsets;

- Information Systems (IS) application servers to handle corporate applications such as Voice over IP (VoIP), Virtual Private Networks (VPNs), firewalling, etc;

- a management console and software for use by the IT staff.

When the network above is first turned on, the WLAN controllers start up, recognize each other, elect or discover a primary, and then automatically discover and initialize the APs. (In some cases the APs require a DHCP server in order to start up.) The RADIUS authentication server is a mandatory part of every enterprise WLAN, and is normally a software package (e.g., Cisco Access Control Server or Funk Steel-Belted RADIUS) running on an enterprise server PC or workstation. The centralized management of the system may be done by means of direct access to the WLAN controllers (e.g., via configuration Web pages supported by the controllers) if there are only a few of them. However, in a mid-sized network this is more commonly accomplished by means of a vendor-supplied software package that is specially designed to manage large-scale WLANs.

8.1.2 Deploying an Enterprise WLAN

As will be clear from the preceding section, an enterprise WLAN is a fairly sophisticated system employing complex equipment. Further, the vagaries of the local RF environment play a large part in the functioning of the installed WLAN. Enterprise WLAN deployment should hence be approached in a systematic and step-by-step fashion, otherwise the final performance of the installed network cannot be guaranteed or even predicted. Note that the deployment of enterprise WLANs (unlike WLANs used in small offices or the home) is now usually done by professional installation contractors, rather than the corporate IT staff themselves. Also, unlike small office or branch-office WLANs, a corporate enterprise WLAN has higher expectations for capacity and security, and a much larger scale; some deployed enterprise WLANs may contain upwards of 15,000 APs and hundreds of controllers, and the investment in structured wiring for AP connectivity alone can be considerable. Thus the WLAN deployment needs to be well-planned and carefully thought out, in order to avoid costly mistakes.

A typical enterprise WLAN deployment takes the following steps:

1. Determine the requirements for coverage, capacity, security, QoS, availability, and cost. Once these are known, the installation contractor can create a proposal or a bid.

2. Perform a site survey to determine interference, fix AP placements, assign RF channels, and set transmit power levels for individual APs.

3. Draw coverage maps and ensure that there are no "holes" in the coverage.

4. Install APs and WLAN infrastructure components. Note that frequently some wired infrastructure components may have to be added as well in order to support the WLAN.

5. Configure and bring up the WLAN.

6. Perform post-installation performance testing to qualify the installation prior to handover to the enterprise IT staff.

7. Fix RF coverage holes (i.e., areas of low or zero signal strength) and tweak channel and power assignments to improve existing coverage.

8. Enable the network for carrying live traffic (i.e., permit users to begin using the installed WLAN).

9. Document the network before turning it over to the IT staff.

10. Once the WLAN has started carrying live traffic, an ongoing maintenance phase begins. During this phase, the network is monitored and problems are diagnosed and fixed. This is typically done by the enterprise IT staff.

11. Upgrade and augment the WLAN as requirements change or improvements are needed.

It is necessary to ensure that requirements placed on the enterprise WLAN are in fact achievable, before trying to plan it or site the APs. The following rules of thumb have been generally found to be useful:

*   no more than five to six WLAN laptop clients per installed AP;

*   no more than five to ten VoIP handsets per installed AP;

*   the total bandwidth required from each AP should be no more than 15 Mb/s;

*   the coverage area of each AP should overlap with all of its immediate neighbors (so that if one AP goes down the coverage loss is small);

*   signal strength throughout the coverage area better than -55 dBm, with higher signal strength available from APs in high-density areas.

A requirement for a higher QoS or availability of service implies a higher density of APs, usually with a lower transmit power setting configured into the APs to prevent cross-interference. Ethernet (Layer 2) virtual LANs (VLANs) should be used between the wired and wireless infrastructure to ensure that traffic requiring different QoS guarantees (e.g., voice vs. data) are kept isolated throughout the infrastructure, and the APs are able to easily recognize high-priority traffic on the wired side. Most enterprise WLAN systems are capable of supporting such configurations.

Determining the channel assignments and transmit power settings for the individual APs in a WLAN is a complicated topic and the subject of much debate. Improperly assigning channels and power can lead to coverage holes, cross-interference, or an overly expensive deployment.

Channels should not overlap in their coverage, and channel assignments should be such that cross-interference is minimized. Ideally, each AP should be on a different channel; however, there are a very limited number of non-overlapping channels (just three in the case of 802.11b/g) that this is usually impossible, and some channel re-use is required. (The task of assigning non-overlapping channels is referred to as the graph coloring problem.) Once the channels have been assigned, transmit power settings need to be adjusted to limit co-channel interference between APs. A high transmit power on an adjacent channel can cause adjacent channel interference, so simply assigning all APs to different channels cannot always solve the problem.

Some WLAN controllers have automatic RF management functions built into them; these controllers will constantly monitor the RF environment (through signals picked up and passed on by the APs they control), and make adjustments in the AP settings to maximize the performance of the complete WLAN. This eases the burden of the installer and the IT staff, as the WLAN can cope with dynamic changes in the environment and surroundings automatically.

8.1.3 Coverage and Capacity

Before actually taking the expensive step of bolting APs into the ceiling and wiring them up, a site survey should be performed, in order to draw up a coverage map and verify that there are no holes in the coverage offered by the APs. Further, the coverage map should be used to estimate the capacity offered to clients by each AP at the limits of its coverage; most AP vendors offer application guidelines that allow installers to estimate the transfer rate as a function of range from their APs in typical indoor environments. This step allows the network planner to see if the bandwidth and QoS requirements can be met as desired, or if changes are necessary to the proposed deployment. An example of such a coverage map (obtained from the AirMagnet Survey tool) is shown in the following figure.

The coverage map is created by performing a site survey, as described later in this chapter. A building floorplan should first be acquired or copied; the map is then drawn on the floorplan, and should indicate:

- the target placements for the APs;

- the target channel assignments for the APs;

- the area covered by each AP, after factoring in antenna patterns for cases where directional antennas are used (such as on the edges of the building floor);

- the location and strength of known interferers and neighboring WLANs;

- the location and approximate size and shape of large metallic objects within the coverage areas.

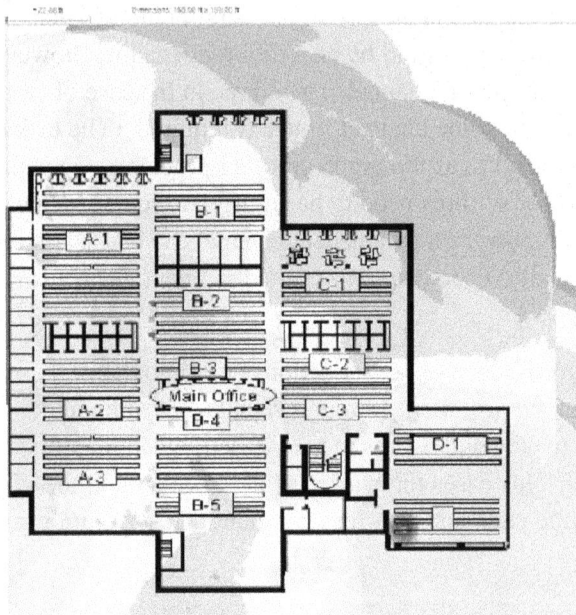

Sample report from AirMagnet Survey

Signal 0 −10 −20 −30 −40 −50 −60 −70 −80 −90 −100
(dBm)

Figure 8.2: State-of-the-art Coverage Map Example
Photo copyright © AirMagnet Inc., provided by courtesy of AirMagnet Inc.

The first thing to be deduced from the coverage map is whether there are likely to be any holes in the coverage (i.e., regions of low or zero signal strength) based on the minimum signal strength criteria established using on the requirements set by enterprise IT staff and the guidelines of the AP vendor. Next, potential problem areas such as shadowing or fading caused by metallic obstacles, interference issues, etc. should be inferred and marked on the map. If active monitoring devices are to be installed as part of the WLAN, the locations of these monitoring devices should be set; ideally these devices should be placed at the limits of the coverage areas of the APs, to detect interference and loss of signal. Finally, the available capacity of the installed WLAN should be calculated (based on the number of clients expected to be served by each AP); at this point, it is possible to determine if the number of APs should be increased or decreased in order to meet the target cost and service level objectives.

The coverage map also serves as a baseline for verifying the installation once it has been completed. If the post-installation tests differ significantly from the predicted coverage, then it is necessary to investigate and determine the causes of the differences. Further, the coverage map serves as a baseline for future network monitoring and upgrade operations.

In this manner, the coverage map acts as documentation for the RF portion of the WLAN infrastructure, and should be retained and updated for the life of the WLAN.

Note that several caveats should be observed when creating or using a coverage map. Firstly, the coverage map is only an estimate; the actual performance of the WLAN will almost certainly be different, though with care and good tools the prediction can be made reasonably close. Secondly, the coverage map usually represents the conditions at the time that it was made. RF conditions, however, vary over time (frequently on an hour-to-hour basis) and thus a static coverage map cannot represent the reality of changing propagation and interference conditions. Finally, it is very difficult to make a three-dimensional coverage map; thus most coverage maps today cover only one floor, in spite of the fact that RF energy propagating between floors has a material effect on WLAN performance.

The details on how a site survey is conducted and used to create a coverage map are given in a subsequent section.

8.1.4 Post-installation Testing

After the enterprise WLAN infrastructure has been planned using a site survey, a coverage map drawn up, and the equipment installed accordingly, it is important to perform a post-installation qualification process. A post-installation qualification procedure can spot problems in coverage, functionality, and performance, all prior to carrying live traffic (and subjecting real users to problems). Examples of what can be exposed during post-install testing are: coverage holes, mutual interference due to excessive AP power, unexpected interference, leakage between floors, misconfigured security servers, etc.

Post-installation testing is usually done with one or more laptops running some sort of off-the-shelf traffic generation software such as Chariot or Iperf. (An unsophisticated installer might simply elect to ping a known network server, or try to open up a Web page with his or her laptop; however, this is not an adequate test.) The laptop runs one side (one "agent") of the traffic generation software; a counterpart agent is run on a PC or server connected to the wired network, and the two agents exchange data to analyze the performance of the end-to-end link. The laptop(s) are then placed at different locations while traffic is running between the agents, preferably at the same locations as the actual users are expected to reside, and the throughput and Received Signal Strength Indicator (RSSI) are noted. Packet latency and error rates can also be observed, if voice applications are expected to be run (and if the laptop can gather such information). The results are then plotted on the same floorplan as the coverage map and compared with the predicted values. If there are significant differences, then problems have been uncovered, and more investigation is required.

The post-installation test process described above can be carried out at low cost with relatively simple equipment – just a laptop or two running software. However, the process is time-consuming, involving manual measurements and recording of data. Also, the amount of data that can be gathered is relatively little; significant pieces of data such as the nature of the interference and time-varying network performance are missed. To avoid this, products such as SiteSpy (from Wireless Valley, now Motorola) and YellowJacket (from Berkeley Varitronics) are available to perform much more comprehensive post-install test. These products have integrated global positioning system (GPS) receivers and mapping software to automatically correlate measurements to locations on the floorplan, as well as spectrum analyzers and profiling tools to do a better job of detecting RF issues. The SiteSpy tools can also tie into RF propagation prediction software to improve the site survey process.

Note that infrastructure companies such as Cisco Systems and Trapeze Networks also provide software packages that enable installers to evaluate wireless networks based on their products.

8.1.5 Monitoring, Diagnosis, and Maintenance

Wired LAN infrastructure requires a limited amount of ongoing monitoring and maintenance to deal with equipment failures or misconfiguration, as well as new requirements and topology changes (confusingly, the IT industry refers to these as MACs – Moves, Adds and Changes). However, WLANs require far more ongoing monitoring, because they are subject to RF environments that can change significantly even when no network infrastructure changes have occurred. For example, a neighboring office might start up a WLAN on the same or nearby WLAN channels, thus increasing the local interference level. Alternatively, large metal objects such as partitions and filing cabinets could be moved when reconfiguring office space, thus creating coverage holes or introducing spatial fading effects. It is therefore essential to perform continuous monitoring of the WLAN; this enables early discovery of problems, and allows diagnosis and maintenance operations to be performed well in advance of user complaints.

Monitoring can be done in two ways: using the capabilities of the APs themselves, or deploying active or passive sensors around the WLAN coverage area. Both methods have pros and cons, and both will be described in more detail later.

8.2 Hot-spots

The term "hot-spot" is generally applied to a place where public WLAN access is available. Access may be free (e.g., coffee shops or some downtown areas) or tariffed (e.g., airports and hotel lobbies). Hot-spots qualify as small- or medium-sized installations; a typical airport hot-spot may deploy up to 100 APs to cover the entire passenger terminal area.

The hot-spot installation process is generally similar to that of an enterprise WLAN, but the infrastructure components used as well as the usage model are somewhat different. A key characteristic of a hot-spot is that the criteria are connectivity and cost, rather than the usual enterprise requirements of performance, security, and manageability. Access security, when implemented, is for the purpose of preventing theft of service, rather than protection of sensitive information.

8.2.1 Anatomy of a Hot-spot

In its simplest form, a hot-spot consists of a single AP connected by a broadband link (such as a business DSL line or a cable modem) to the Internet; this has no access security and no billing, but merely provides a means for an unspecified number of users to gain access to the Internet. A more typical hot-spot is a network of APs distributed in a building, backed by a specialized hot-spot controller (gateway) plus a local authentication and accounting server, with a T1 or fractional-T1 line to the hot-spot service provider's Point of Presence (PoP) in the metro area; the PoP in turn connects to the service provider's internal backbone network and thereafter to the Internet. A large hot-spot (e.g., in a large airport or a convention center) may have a more extensive installation, such as multiple WLAN controllers, a monitoring and diagnostic system, a local Web and fileserver cache for frequently accessed content, and a proxy server. A typical hot-spot setup is shown in the following figure.

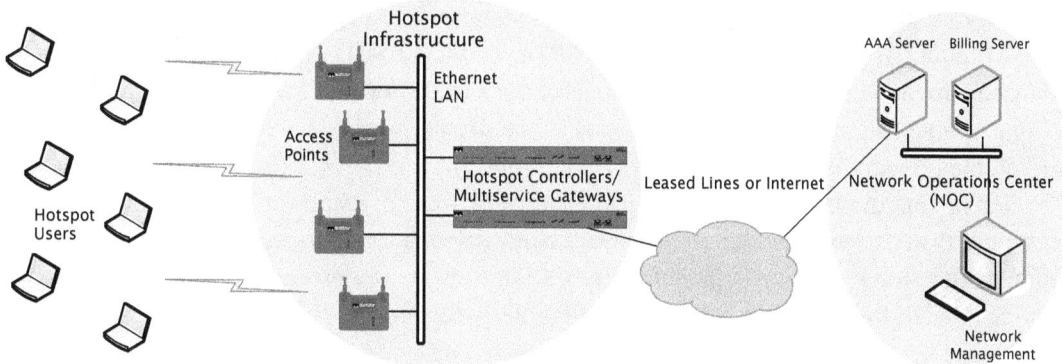

Figure 8.3: Typical Hot-spot Setup

In the above figure, the APs are deployed around the hot-spot venue for maximum coverage, and then connected to the controllers via the private wired LAN infrastructure backing the WLAN controllers.

It is unusual for any QoS, bandwidth, service availability, or other guarantees to be provided to the user by the hot-spot service provider; service is strictly on a best-efforts basis, especially as the WLAN channel in a given area is shared among all the users in that area. Hot-spots belonging to telecommunications service providers that offer a nationwide subscriber data access plan may have such guarantees; in this case they are treated more like an enterprise WLAN than a best-efforts hot-spot.

8.2.2 Hot-spot Deployment Challenges

Hot-spot deployment is much more of a challenge than the usual enterprise deployment, because the hot-spot service provider typically does not control the building and environment in which the deployment takes place. In fact, in environments such as airports there are security and interference concerns that may greatly limit the ability of the service provider to ensure uninterrupted coverage. Also, there are less constraints on the user base (e.g., in terms of what applications are run or what types of laptops are used), and thus ensuring that all users can gain access is more difficult. As a consequence, there may be severe constraints on equipment types, services offered, AP placement, cabling, power output, operating channels, security schemes, and so on. Today, hot-spot equipment is custom-built with the necessary capabilities required by service providers, rather than merely using off-the-shelf enterprise infrastructure components.

8.3 The Site Survey

Some kind of site survey should (and usually is) be performed prior to any enterprise-class WLAN deployment; apart from the rigorous planning needed to avoid expensive mistakes, it is essential to understand the RF environment to some degree before attempting to use it for access. Modern WLAN infrastructure, with "RF-aware" WLAN controllers and specialized RF management software, reduces the extent of the site survey but does not entirely eliminate it. For example, if a strong interferer such as a 2.4 GHz long-range wireless video system is present, then no amount of RF management will prevent a WLAN in that area from being impacted or shut down entirely; the interferer must be moved, its power must be reduced, or else the WLAN has to switch to the 5 GHz band – which, of course, changes all of the requirements on AP placement and client types. The site survey process provides advance warning of such impending catastrophes, to enable better project planning, more accurate costing, and less surprises during network activation.

Attention to detail during a site survey is essential, even though it is time-consuming to gather all the information required. Small details, such as a Bluetooth headset system on a receptionist's desk, can make a big difference in a crowded office area. (Bluetooth uses the same 2.4 GHz ISM frequency band commonly used by WLANs, but its frequency-hopping physical (PHY) layer can shut down the surrounding wireless network.) A systematic approach is also essential, otherwise a lot of wasted effort can result. Good tools are also important in order to avoid untrustworthy results.

8.3.1 Site Survey Objectives

The site survey process seeks to create the coverage map referred to above (see Section 8.1.3). A good site survey also encompasses planning for AP placement that satisfies the coverage and capacity requirements; these requirements should be drawn up prior to starting on any survey work, and the site survey should be carried out with the requirements in mind.

The output of the site survey is a report and coverage map detailing the following pieces of information:

- The building floor plan, with prospective AP locations marked on it.

- The coverage obtained by the APs when placed in those locations (channel assignment is usually done later, when the final placements have been determined).

- The locations of known interferers, with some indication of their type, and an estimate of their coverage relative to the predicted signal strengths of the APs.

- The locations of any pre-existing APs that may be present, with their measured signal strength contours.

- Reports of any issues found that could affect either coverage or AP placement (e.g., large obstacles, concrete walls where cables are difficult to run, leakage across floors, etc.)

From the coverage map and reports, a network planner or IT staffer can determine the number of APs required to cover the desired area with the target level of signal strength. Further, the coverage map will also provide indications as to the best AP placement, the channel selections needed in order to minimize interference between APs, the estimated bandwidth available to users in different regions, places where extra APs may be needed (e.g., conference rooms), and potential trouble spots.

The output of the site survey enables network managers to "plan the air", and deduce the end-user's perceived results before the first end-user starts using the WLAN. Note, however, that some limitations exist in even the most comprehensive site-survey process. The survey and planning results are not as clear-cut as those for a wired (switched) Ethernet network: bandwidth is still shared, environmental conditions still change over time, and denial-of-service attacks can still occur.

8.3.2 Tools for the Site Survey

The minimum toolset used in a site survey is simply a stand-alone AP (or two), configured to use a specific channel, and a laptop with a WLAN NIC (Network Interface Card). To this the systems integrator adds a ladder, some extension cords and power supplies, and a tape measure. The software required is minimal, too – the APs need none, and the laptop NIC usually already comes with some kind of software utility that displays RSSI and noise.

The above is a minimal setup, and the site survey report produced is likewise rather spartan. For example, it is difficult to do anything more about interference than deduce that it *might* be present; without a spectrum analyzer and a directional antenna, it is not possible to determine either the type of interference or its location. There is also no way to run any kind of actual traffic tests, as the APs are not traffic generators and not wired into any kind of traffic sources. Instead, the installer must depend on the AP beacons exclusively, which provides only RSSI indications using the 1 Mb/s data rates at which the beacons are transmitted. Determining what happens at higher data rates and traffic loads is out of the question; however, an environment that appears reasonably benign when measured using 1 Mb/s beacons transmitted at 10 per second may prove to be quite otherwise when trying to transfer thousands of packets per second at 54 Mb/s. Finally, laptops take a relatively long time to average out RSSI variations and local noise levels, so the number of measurement points that can be taken is quite low, usually no more 1 per room or cubicle.

A more comprehensive site survey toolkit would include one or more of the following:

- A low-cost spectrum analyzer with an optional directional antenna (e.g., the AirMagnet Spectrum Analyzer or the Fluke Networks AnalyzeAir), to pick up interference, characterize it, and localize it to a specific location.

- A handheld PDA-type device (such as the YellowJacket from Berkeley Varitronics, see Figure 2.11) that is specifically dedicated for site surveys and can quickly indicate not only signal strength and spectra but also analyze multiple channels, detect the presence of existing WLANs, and automatically record results into non-volatile storage.

- An automatic position reporting and recording system, optionally linked to the handheld, that automates the task of marking locations at which samples are taken; such systems may either use GPS or some other location tracking system.

- A handheld traffic generator and analyzer (TGA), such as SiteScanner from Motorola (formerly Wireless Valley), that generates actual data traffic to the test APs and records the results, allowing measurements at different PHY rates using live traffic.

- A laptop with WLAN sniffer software (such as WildPackets Airopeek or AirMagnet Laptop) for passively monitoring and identifying adjacent WLANs in operation, as well as observing client/AP conversations and traffic density.

In addition, various vendors supply software tools that can be used to streamline and speed up site surveys. Such software includes architectural drawing packages (e.g., AutoCAD) that can be used to represent the building database (i.e., floorplans, materials, furniture, etc.), and dedicated site survey support software to improve data gathering and recording of issues and automatically produce a report at the end of the process. Proper tools such as YellowJacket and AirMagnet Survey can make the site survey process much easier and more accurate.

8.3.3 Site Survey Process

Most site survey processes follow the same general outline, as shown in the following figure:

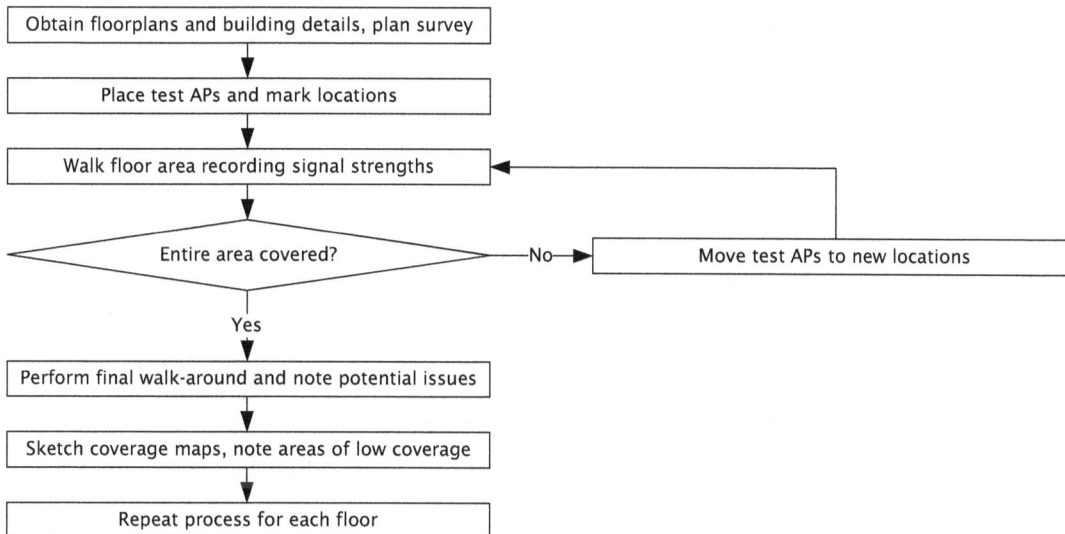

Figure 8.4: Site Survey Process

The steps followed by this "typical" site survey process are:

Step 1: Obtain floorplans and other building information.

It is not possible to perform a systematic site survey without a floorplan of some kind; at the minimum, one is needed to plot measurements, record candidate AP placements, etc. Unfortunately it is often difficult to obtain such floorplans, especially in older buildings where the floorplans may have been misplaced or exist only as blueprints. In such situations the installer has little alternative but to draw one from scratch.

Step 2: Place test AP(s) and mark locations on the floorplan.

A set of test APs must be placed at candidate locations to serve as signal sources. These test APs should be as similar as possible to the make/model of APs that will be installed in the real WLAN. AP placements are critical to the final results of the site survey. However, until the survey is partially done it is not possible to tell whether the initial placements were good choices. As a consequence, this important step is usually rather hit-or-miss, and is based on educated guessing by the installer; experience counts for a lot in such cases. In general it is better to start at one end of the building, dividing up the space to be surveyed into relatively small regions, and work towards the other end.

Step 3: Walk around recording signal strengths and other information on the floorplan.

This is the most time-consuming step in the process; the installer must capture signal strengths, interference, existing WLANs, etc. in the areas surrounding the test APs.

Step 4: Move test AP(s) to new location(s) and repeat walk around and recording process.

It is unlikely that enough test APs will be available to enable the entire site to be characterized in one pass; thus it will be necessary to cover the area progressively, shifting the APs after each pass. It is essential to place the test APs such that their coverage areas overlap with the previously surveyed area, so there are no blind spots.

Step 5: Perform a final walk around, noting potential problems on the floorplan.

Examples of such problems are: large obstacles of metal or concrete; thick concrete or brick walls which will make cables difficult to run; potential RF screening or shadowing zones, such as elevator shafts; and intermittent interfering devices (e.g., a wireless security camera that is turned off during the survey, but could be turned on again later in the day).

Step 6: Create coverage maps and identify areas of low signal strength.

The installer should plan for additional APs to "fill in" regions that are not adequately covered, on the basis of the signal strength measurements.

Step 7: Repeat for all floors on which WLANs need to be installed.

It is also a good idea to check for inter-floor interference (e.g., propagation through thin concrete floors or diffraction via exterior windows) as well at this time, though this is a laborious and time-consuming step that is often omitted. Luckily, RF propagation through concrete floors is relatively poor and can be ignored in most cases.

8.3.4 Reporting

Reporting is as important a component of the site survey as making the actual measurements; a comprehensive report should be produced at the end of the process, which will serve as the deliverable from the installer to the customer. The reporting requirements vary by the tools used, the size of the installation, and the methodology followed by the installer. At a minimum, the site survey report should contain the coverage maps resulting from the candidate AP placements, as well as the other data gathered during the process such as the potential trouble spots and the level of interference.

The report should be preserved and updated periodically; it not only serves as essential documentation for pre-install activity, but also as a baseline reference for post-install diagnostics and monitoring. For example, if problems crop up later, test APs can be placed at the original locations and the survey process repeated to determine whether the coverage has changed. This can save a lot of unnecessary debugging of network infrastructure (e.g., clients suddenly unable to connect) when RF environment changes have occurred.

8.3.5 Pitfalls

The site survey process attempts to characterize a complicated phenomenon (indoor RF propagation) using a relatively small number of data points, and is therefore subject to a number of potential issues. These should be kept in mind when performing the survey and interpreting the results.

Firstly, the placements of test APs significantly affect the quality of the results. As previously noted, initial placements are based on installer guesswork, experience, and instinct. Repeating the site survey for different test AP placements can be very burdensome, thus if an initial placement is barely adequate or "tweakable" there is frequently no effort put into changing the placements and redoing the survey. This hit-or-miss approach definitely does not provide an optimal solution – for example, the output of the site survey may indicate that many more APs are required than originally expected.

Secondly, the survey process takes a long time and a great deal of manual effort. This produces significant possibilities for error, as well as problems created by installers taking shortcuts or skipping measurements.

Another issue is that the site survey is usually a one-time snapshot of conditions. (It is quite laborious doing a *single* site survey; requiring an installer to do several over the course of a day or a week is quite unreasonable!) However, the actual indoor RF environment changes on an hour-by-hour and day-by-day basis, according to workflow patterns and changes in the surroundings. Thus a considerable amount of margin has to be built into the results in order to deal with the variations.

Also, it is difficult to convert coverage and signal strength measurements made during the site survey process into true capacity and mutual interference figures; the installer or tool has to estimate these figures based on empirical data supplied by the AP vendor as well as experience. This is because, as noted above, the test APs used as signal sources are only emitting beacons, not handling actual traffic. Beacons arrive at a slow rate (10 per second) and fixed bit rate (1 Mb/s), unlike regular data traffic which may produce thousands of packets per second at a variety of bit rates. Therefore, interference with the actual data traffic may not be found during the site survey, but can manifest itself later, when the network "goes live". (Some tools – e.g., AirMagnet Survey – can run data traffic to the test APs.)

To some extent, the above issues can be mitigated by a three-step process:

1. First, performing a comprehensive site survey to get a rough idea of the "lay of the land".

2. Second, over provisioning the system by some factor, to provide reserves of channel capacity and transmit power that can be used to overcome undetected interference and shadowing effects.

3. Third, enabling automatic RF management functions in the WLAN controllers and switches to dynamically set channels and transmit power, thereby utilizing the reserve capacity to maintain continuous availability and high performance.

The last step is possible as a result of the much more capable and powerful RF management functions available in enterprise-class WLAN controllers today. Such controllers are capable of automatically and continuously receiving, analyzing, and integrating signal, noise, and interference measurements from their connected APs; forming an assessment of channel conditions and interference caused to or by nearby devices; and setting AP channels and power to maintain the desired traffic rates while minimizing mutual interference. In some cases, the WLAN controllers are even capable of instructing the client laptops and handhelds to increase or reduce power in order to minimize the effects of interference.

8.4 Propagation Analysis and Prediction

A (potentially) much more accurate method of determining coverage, bandwidth and other parameters uses complex RF propagation modeling software to simulate and analyze an indoor RF environment, and predict the signal strength contours at all points within the environment. From the signal strength contours and the characteristics of the equipment to be installed, the path loss, throughput, error rate, etc. can be deduced. Once the necessary amount of building data has been gathered and input to the software program, this is a far faster method of determining optimum AP placement, as it does not involve trial placements of actual APs followed by tedious walking around.

8.4.1 Indoor Wireless Propagation

Propagation of RF signals is basically identical to the propagation of light, with the significant exception that the wavelength of interest is much larger; thus metallic objects smaller than a few centimeters in size are effectively "invisible" to RF energy produced by WLANs at 2.4 and 5 GHz, and materials that are opaque to light allow RF to pass through them. Further, the propagation medium does not change in relative density (in terms of the dielectric constant ε) very much over the short distances involved in indoor environments, and so refraction is not usually a factor. With these exceptions in mind, the familiar optical principles of straight-line propagation, reflection, diffraction, etc. apply.

Four key effects control RF propagation in an indoor environment:

1. *Attenuation (absorption)*: Walls, partitions, floors, ceilings, and other non-metallic objects – including humans! – attenuate radio waves passing through them. In extreme cases, virtually all of the RF energy may be absorbed, in which case the region behind the object is in an RF shadow.

2. *Reflection*: Large metallic objects, with dimensions substantially greater than one wavelength, reflect RF energy impinging on them according to the standard principle for optical waves (i.e., the angle of reflection is equal to the angle of incidence.) Reflection from metallic objects also causes RF shadows.

3. *Interference*: If two or more waves arrive at the same point in space but take different paths, and hence have different path lengths, then constructive and destructive interference occurs. In the case of RF, this is usually referred to as fast fading.

4. *Diffraction*: Large metallic objects with distinct edges, such as metal sheets or furniture, cause diffraction at their edges, and enable propagation into areas that would otherwise be in RF shadows.

The following figure illustrates the various mechanisms underlying RF propagation in an indoor environment. See Figure 3.4 as well.

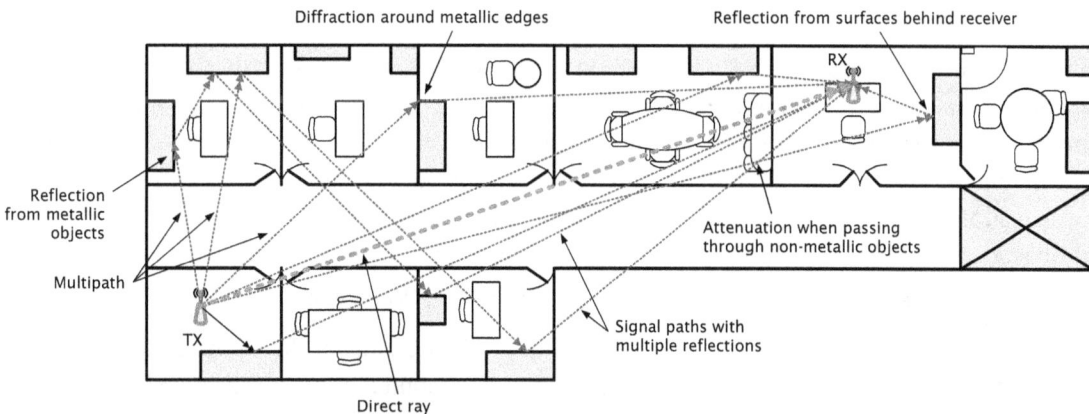

Figure 8.5: Indoor Propagation

It is convenient to express the path between transmitter and receiver, which has a particular set of properties that affect signals passing from the former to the latter, as an RF "channel" (in the same sense as a waterway). These properties are determined by the propagation effects imposed on transmitted signals before they get to the receiver. As the indoor environment is very complex and not easy to calculate exactly, statistical methods are usually used to model the channel and estimate its effects upon RF signals. The channel is referred to either as Ricean or Rayleigh, depending on the statistical distribution of amplitudes in the signals arriving from the transmitter at various points in the environment.

In empirical terms, a Ricean channel generally has a strong line-of-sight component (i.e., the bulk of the RF energy propagates in a straight line from transmitter to receiver). A Rayleigh channel, on the other hand, has the bulk of the energy arriving along non-line-of-sight paths.

For relatively low data rate PHY layers such as 802.11a, 802.11b, and 802.11g, the distinction between Ricean and Rayleigh channels is not very important. However, for 802.11n, this makes a significant difference, as we will see in the next chapter.

8.4.2 Propagation Models

A propagation model is the term given to a statistical model of a channel between any two points, in terms of Ricean or Rayleigh statistics. Due to the complexity of the indoor environment, however, these models are frequently implemented as computer programs rather than equations. Two kinds of propagation models have been generally used: parametric models, which express the channel properties in the frequency domain, and ray-tracing models, which operate in the spatial domain. The most common modeling and simulation approach used for the indoor environments that WLANs are concerned with is ray-tracing, as this approach is best able to deal with the complexity of the environment.

8.4.3 Propagation Simulation

Propagation simulation originally focused on implementing models (usually parametric) for satellite and cellular communications, but now extends to indoor propagation – usually ray-tracing, as described previously. Such propagation simulation is fairly complex because the indoor environment is full of artifacts (walls, ceilings, doors, furniture) that affect RF propagation. However, the use of powerful computers makes it possible to simulate the propagation accurately within quite large indoor areas. Ray-tracing, borrowed from computer graphics, is the principal means of performing indoor RF propagation simulation today.

The ray-tracing method is conceptually very simple. The features of the environment (doors, walls, etc.) are represented to scale on a grid within a computer, resembling an architectural floorplan, but referencing the RF properties of the various elements. A simulated RF "source" is placed at some desired location. "Rays" are then drawn in all directions from the RF source, representing electromagnetic waves propagating linearly outwards from the source with a given signal strength. Where the rays strike elements of the environment, the laws of propagation (i.e., reflection, attenuation, diffraction, etc.) are applied to determine the magnitude and phase of the resulting transmitted and reflected rays. If two or more rays intersect, then interference calculations are made to determine the resultant signal strength. The process is carried out until some desired degree of resolution is reached; plotting the signal strength at each point on the floorplan then gives the simulated propagation of RF from the simulated source. Using the principle of superposition, the procedure can be repeated for any number of sources at different locations until a complete picture of the RF signal strengths within the environment is obtained. Figure 8.6 shows a simplified view of this process.

The ray-tracing method is computationally intensive but is very powerful. If the dimensions and RF properties of the objects within the environment are known, as well as the

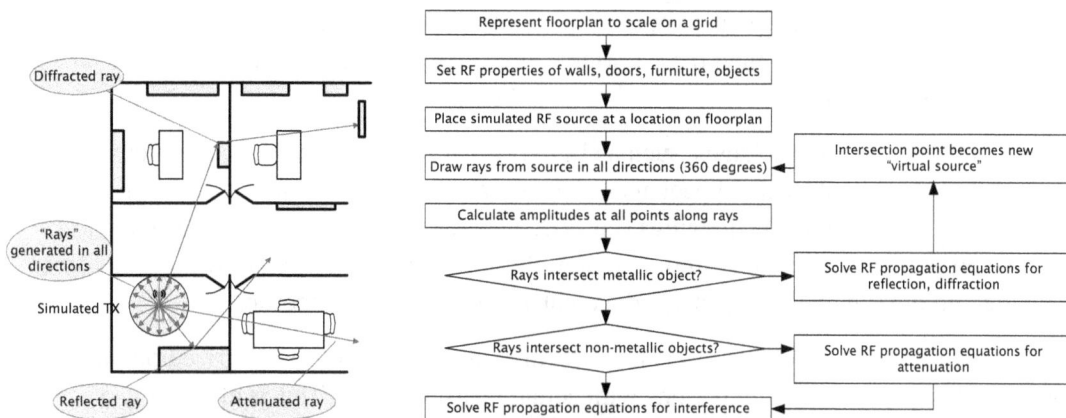

Figure 8.6: Ray-Tracing Simulation Process (Simplified)

properties of the source, then the RF field strength can be plotted very accurately at any point. Experimental comparisons between the ray-tracing method and actual propagation measurements show very good correlation, and it is now the de facto method for indoor propagation simulation.

8.4.4 The Prediction Process

With ray-tracing simulation, it is possible to bypass the manual site survey process and directly predict the coverage and throughput available from a given AP placement. This type of prediction process is as follows:

- The building floorplan and material characteristics (i.e., the RF properties of walls, furniture, etc.) are entered into the simulator.

- The RF characteristics – transmit power, antenna radiation pattern, etc. – of the equipment (APs) are also entered.

- A set of candidate AP placements are made on the floorplan.

- The simulator then takes over, performs a ray-tracing simulation, and plots the coverage (in terms of signal strength contours) on the floorplan. Once the signal strength is known, the simulator may even deduce the available throughput at various points based on the characteristics of some selected WLAN receiver.

- The coverage and throughput contours are manually inspected. If the coverage is unacceptable, the AP placements can be changed and the simulation re-run immediately.

The prediction process is far faster than the manual site-survey, provided that the building and equipment characteristics are known in advance. Further, it is possible to perform many "what-if" scenarios and arrive at an optimal placement. Obviously, this is a much simpler and less labor-intensive process than the traditional site survey – if accurate and complete data on the building is available.

Several commercial SW packages, such as LANPlanner from Motorola Inc., implement this process. The more sophisticated packages support various features, such as automatic entry of floorplans from AutoCAD drawings (i.e., DXF format files), a large materials database with RF properties of common building materials, a floorplan editor to allow users to place furniture and other metallic objects, and a database of APs with properties.

One extension to the above process is to perform a prediction of coverage based on known data, and then to refine the predictions with actual measured data. This is effectively a blending of the propagation modeling and the site survey processes. The floorplan and materials are entered first, the propagation is modeled, and initial predictions of coverage made. An actual AP or signal source is then placed at a target location and a special receiver is used to record signal strengths at some points around the coverage area. These data points are compared with the predicted values from the simulation, and the differences between measurements and prediction are used to refine the propagation modeling and compensate for errors. This allows a more accurate result, but without all the manual labor of the site survey. Tools such as InFielder from Motorola Inc. assist here.

8.4.5 Modeling Equipment Characteristics

As the purpose of the installation process is to determine the optimal placement of APs, only the APs really need to be characterized. (While the RF characteristics of the client cards play a significant role in the actual end-user experience, the installer has little control over this; all he or she can do is to place the APs at optimal locations to assure the desired signal strength and coverage.) For the purposes of propagation modeling, APs can be characterized by three factors:

1. *The total radiated power:* This is the transmit power integrated over three dimensions (i.e., the total power output of the transmitter minus the power lost in the antenna and cabling).

2. *The total isotropic sensitivity*: This is the sensitivity of the AP as integrated over three dimensions (i.e., the sensitivity of receiver divided by the efficiency of the antenna and cabling).

3. *The antenna radiation pattern.*

If these three factors are known, then the coverage (receive and transmit) of the AP can be predicted using the ray-tracing simulation process.

Unfortunately, most vendors do not publish any of the above characteristics. However, they can be approximated; further, for most purposes it is only necessary to model the transmit characteristics of the APs. The receive coverage is assumed to be about equal to the transmit coverage, which is true for most well-designed APs. In addition, the total radiated power can be approximated as being equal to the transmit power of the AP (this assumes losses in the radiating system are negligible, which in most cases is true). This leaves the antenna radiation pattern as the unknown factor. If standard vertical antennas are used on the APs, then the antenna radiation pattern can be assumed to be the typical doughnut shape of a vertical dipole. On the other hand, if a directional antenna such as a patch is used, then the radiation pattern is no longer a doughnut, but has lobes (regions of higher signal strength) and nulls (regions of lower signal strength) in various radial directions. These lobes and nulls can be predicted, with a bit of difficulty, from known antenna radiation patterns. These two can be plugged into the propagation modeling software, and the resulting coverage contours plotted.

Fortunately for the installer using commercial propagation modeling software, these characteristics have already been incorporated into the software for many commonly available APs. All that the installer needs to do is to select the appropriate AP from a list and orient it on the on-screen floorplan as desired. The software will then consult its database of equipment properties and obtain the information necessary.

8.4.6 Limitations and Caveats

The propagation modeling process can produce results that are very close to reality, but the biggest limitation is the need for complete and accurate entry of environmental data. Without complete knowledge of the indoor space, producing a truly accurate picture of the RF channel is difficult or impossible.

"Complete" here is to be taken literally; every large metallic object needs to be input (heating ducts, elevator shafts, cubicle walls, etc.) and the RF properties of every wall, door, and window must be entered as well. However, the architectural drawings are often not available, or are not in a form that is readily acceptable to the software. (A stack of blueprints makes for a laborious and tedious process of conversion into a vector drawing, such as with AutoCAD.) Further, even if such drawings were available, the actual building very frequently diverges from the architectural drawings, thanks to changes and architectural license taken during the construction process. Further, the materials composing the floors, walls, and ceilings are often not known; even if they are, the RF properties of the materials may not be known. Details such as the furniture play a significant role in the propagation, but these materials and dimensions are even less well known than the walls and partitions.

Another limitation is that the surroundings can play a substantial role in the RF propagation, but is typically difficult or impossible to model. For example, large glass windows are transparent to RF; a concrete wall just outside the windows can therefore reflect RF back into the indoor space, substantially changing the field strength pattern. Predicting interference from neighboring areas is particularly difficult.

Finally, as has been noted above, the characteristics of the equipment (APs, etc.) are not straightforward to include, as they are not usually available from the vendor and not easily measured without complex equipment. Apart from the variations in equipment RF characteristics due manufacturing tolerances, there is also an impact due to cabling (e.g., the angle at which cables are run to and from the APs) and the proximity of surrounding metallic objects, which will alter the radiation pattern of the APs.

All of these effects make propagation modeling considerably less accurate than would normally be expected. Fortunately, the level of accuracy needed for arriving at a workable placement of APs is relatively low; with a moderate safety margin, it is possible to obtain fairly good results even in the absence of all the comprehensive information regarding the indoor space. The ability of enterprise WLAN controllers to "manage" the RF environment also simplifies the task; small errors in the modeling process can be masked by changing the transmit power of selected APs up or down to compensate.

8.5 Maintenance and Monitoring

Wired enterprise LANs require continuous monitoring and maintenance for proper operation; WLANs in the enterprise are not exempt from this requirement either. However, WLANs have a further complexity in that they are subject to changes in the surroundings and the interior physical environment, which makes monitoring even more important. Some examples of changes in the indoor environment that could significantly alter the operation of a WLAN are:

- new interferers (e.g., a newly installed but leaky microwave oven);

- malicious intrusion from outside;

- changes in propagation conditions causing coverage loss, such as metallic furniture being moved;

- the installation of neighboring WLANs, causing an increase in the channel congestion;

- an increase in the number of clients; unlike wired LANs, where the number of clients is limited to the number of physical ports, a WLAN can see arbitrary increases in client counts as users bring in laptops and handheld devices.

Such changes can cause significant adverse impact on the operation of the WLAN as originally installed, and the WLAN configuration may have to be modified to cope with these changes and restore the same level of service formerly provided to the users.

8.5.1 Monitoring and Maintenance Tools

As mentioned previously, two kinds of tools are utilized for WLAN monitoring and maintenance. Firstly, the APs (and WLAN controllers) themselves contain quite extensive built-in statistics and data gathering facilities, that function even as the APs are operating to support clients. In addition, several vendors offer dedicated diagnostic tools specifically designed to address issues in enterprise WLANs. In many respects these are complementary approaches; the built-in tools within the WLAN infrastructure can alert the IT staff to problems, and the dedicated tools can be used to localize and diagnose these problems and verify solutions.

The built-in monitoring capabilities within virtually all enterprise-class APs represent the simplest and cheapest way of performing continuous monitoring of installed WLANs. Considerable passive surveillance and monitoring functions can be performed using these facilities, which can track the level of interference and noise surrounding each AP, scan channels to find WLAN devices in the neighborhood, monitor signals received from neighboring APs and clients belonging to the same WLAN, monitor signals from APs and clients that are not part of the same WLAN (sometimes referred to as "rogues"), and detect malicious attacks or intrusion attempts. If problems are suspected with clients, the APs can perform simple RF tests on the clients by exchanging packets with them and tracking the results. WLAN controller-based systems are particularly effective at monitoring, as they can integrate information received from multiple APs and report it up to the management console as a network-wide report. Further, these monitoring functions can integrate into large, widely used enterprise network management platforms (such as OpenView from Hewlett Packard) and provide the IT staff with a picture of the wired and wireless network as a unified whole.

The advantages of having the monitoring functions built into APs are:

- low cost

- simplicity

- reduced cabling and infrastructure complexity

- less management overhead (less devices to manage and maintain)

- easier setup

- information can be shared between network management and network monitoring.

The sharing of information between network management and network monitoring is a powerful argument in favor of building monitoring functions directly into the WLAN infrastructure. For example, clients can be identified as legitimate by the WLAN controller based on the security credentials negotiated when they connect, and this information can be used to automatically screen out valid clients when checking for rogues and intruders. This can greatly reduce the burden on the IT staff.

Dedicated diagnostic tools typically comprise the same equipment as used in site surveys: laptops with "sniffer" software, spectrum analyzers, handheld PDA-based signal monitors, etc. When a problem is detected, these tools are used to localize and identify the nature of the problem, and diagnose the root cause. For example, a sniffer can be used to scan for intrusion attempts or denial-of-service attacks, or WLANs that have started up in adjacent offices or buildings. In some cases a "mini-site-survey" can be performed using the tools, to systematically locate and diagnose the issue. (It is useful to have the results of the original site survey available for comparison, so that large changes in properties can be quickly identified.)

8.5.2 Active Monitoring

Companies such as AirMagnet also provide dedicated monitoring functions using a hardware monitoring architecture. In these products, wireless monitoring "sensors" are deployed around the WLAN coverage area, and connected to the wired infrastructure; these devices are independent of the APs and WLAN controllers, and are installed and operated as a separate subsystem. The sensors can pick up and track all the WLAN signals in their surrounding area; sampling techniques allow them to track multiple channels concurrently (though not simultaneously, unless special radios are used). The sensors then feed information to a management server that aggregates and consolidates all the information, after which a management console can be used by the IT staff or network administrator to inspect and analyze the data. The sensors can operate in remote offices as well as locally, thus enabling an entire corporate-wide network to be managed as a unit.

Such a distributed system can monitor for many problems:

- malfunctioning or misconfigured APs and clients,
- rogue APs or clients,
- excessive noise and interference levels,
- malicious attacks and intrusions,
- APs or clients suffering excessive traffic loss due to weak signals,
- mutual interference between WLAN devices,
- excessive collision levels and channel overload.

Active monitoring offers several advantages when compared to building similar functionality into the APs themselves. The sensors are dedicated, and hence can monitor continuously (an AP cannot monitor when it is transmitting, and vice versa). Also, they can switch rapidly from channel to channel, or even monitor multiple channels concurrently; an AP must stay on one channel or risk dropping all its associated clients. These systems can detect a much larger range of issues, as they typically use special radios backed by powerful analysis software. The sensors can be placed in known problem areas, thus eliminating the need to choose between the best sensing locations vs. the optimum AP placements. Also, converting all APs into

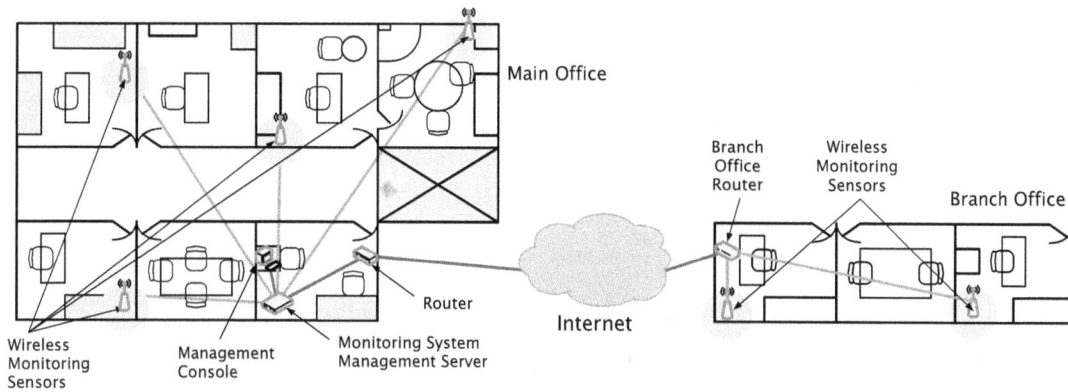

Figure 8.7: Active Monitoring Setup

sensors makes them more complex and costly; usually, a much lower number of sensors is required, as compared to the number of APs. Finally, dedicated sensors can detect issues with the APs themselves; for example, if an AP's radio is malfunctioning, then a sensor can detect this, but the AP cannot diagnose itself.

8.5.3 Smart WLANs

Recently, enterprise WLAN vendors have started building a great deal of intelligence into the control software within their products. These products already comprise a central controller (or array of controllers) that manage a large number of APs; thus they are ideally set up for an intelligent, centralized approach to management of the entire WLAN installation, rather than forcing the enterprise IT staff to deal with single APs at a time. For example, the controller can take over functions such as channel assignment, power control, and interference mitigation for the entire WLAN, using information gathered from the whole installation, and coordinating the activities of all the APs at one time. A schematic diagram of such a "smart WLAN" is shown in the following figure.

In a smart WLAN, APs listen to each other, to their associated clients, to external signal sources (such as adjacent APs and clients), and even to noise and interference on the channel. In some vendors' equipment, APs can be configured to spend a small fraction of their operating time (typically under 1%) scanning for activity on channels other than their assigned channel, gathering still more information. The central controller then receives all this information and makes decisions as to RF channel assignment, transmit power levels, client association limits, traffic load distributions, QoS parameters, etc.

In particular, channel assignment and transmit power level control become automatic and adaptive, and do not require any intervention on the part of the IT staff; the WLAN controller assigns channels such that adjacent APs do not interfere with each other, and reduces transmit

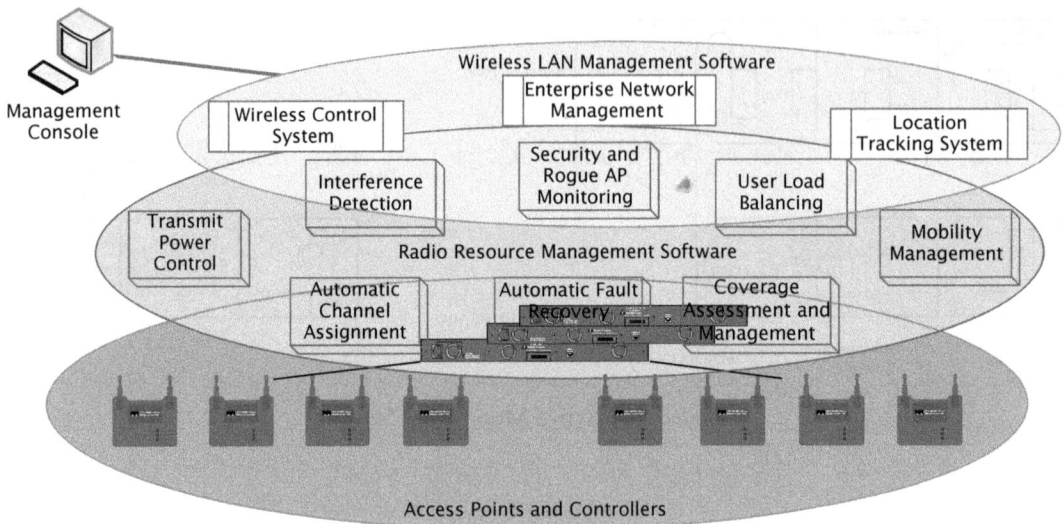

Figure 8.8: Radio Resource Management Architecture

power to ensure that distant APs on the same channel cannot hear each other. If an interferer appears on a channel, the central control system can deal with it by several means: increase transmit power on the affected APs (and, in some cases, the clients as well) to overcome the interference, switch channels to bypass it, or report it to the system administrator for manual action.

Several advanced features are also possible in such an automatically managed system. One of these is load balancing: the WLAN control system can recognize when an AP is overloaded, find adjacent APs that may be capable of taking over the load, and force some of the clients on the overloaded APs to re-associate to the adjacent APs. Another capability is coverage hole detection: if a client is received at low signal level on all the neighboring APs, the system can notify the system administrator that a coverage hole exists at that location. Self-healing to cope with equipment failures is another advanced capability. If an AP goes down, or propagation changes reduce its coverage area drastically, then the controller can automatically readjust the power on the adjacent APs to compensate (i.e., fill in the resulting coverage hole).

These smart WLAN features cannot completely eliminate the need for site surveys and pre-installation planning, but they can help to mitigate issues due to errors or incomplete data (e.g., interference that occurs only sporadically). With enterprise WLANs offering such features, the installer can perform a shorter, less accurate site survey, and then over-provision (install more APs than necessary) by a small amount. Once the intelligent radio resource management functions are turned on, the over provisioning is converted automatically to reserve capacity and power, which can then be used to overcome unexpected issues or changes in the environment.

Testing MIMO Systems

The IEEE 802.11n draft standard (scheduled to be ratified in 2008) uses advanced Multiple Input Multiple Output (MIMO) radio techniques, using two or more simultaneously active antennas combined with two or more transmitter and receiver channels. MIMO promises to provide order-of-magnitude increases in physical (PHY) data rates along with increased resistance to interference and greater effective range (distance between the transmitter and receiver). However, these techniques are particularly difficult to test, as they are both complex and highly sensitive to the RF environment in which they are deployed. This chapter covers some of the special needs and approaches for testing MIMO devices and systems.

Note that as much of this technology is just being developed, and the whole area of MIMO in wireless LANs (WLANs) is still very much in its infancy, many of the test techniques and approaches are still under research and development. This chapter should, however, arm the prospective test engineer with enough background to get a start on the MIMO testing process. Before diving into test techniques, however, we will take a reasonably detailed look at what MIMO is and how it works.

9.1 What is MIMO?

MIMO is the term given to a technique whereby multiple antennas, transmitters, and receivers are exploited in an RF multipath environment (see Chapter 3) to provide a radio link with increased information capacity, improved interference suppression, greater range, and higher fading resistance. The term "MIMO" encompasses a number of different techniques, ranging from relatively simple smart antenna systems to complicated space-division-multiplexing and multi-user detection (MUD) arrangements. For the purposes of IEEE 802.11 WLANs, we are concerned mainly with the use of MIMO techniques to create multiple substreams of data between the same transmitter/receiver pair, thereby multiplying the capacity of the link between the transmitter and receiver. MIMO involves a tremendous amount of highly complicated digital signal processing (DSP); we will not go into the details here (the reader is referred to the many good books on the topic, such as "Space–Time Wireless Channels" by Durgin), but instead provide a brief overview to understand how MIMO works and its effects on both WLAN equipment and test procedures.

9.1.1 Putting Scattering to Work

As noted in previous chapters, an indoor environment is usually rich in metallic obstacles that reflect or diffract radio waves (i.e., scatter them); such obstacles are referred to as scatterers. To recapitulate, scatterers cause RF signals originating from a source (transmitter) to take multiple paths, of different lengths, before they arrive at a common destination (the receiver). The different paths that are traversed by RF energy originating from a single source and impinging on a single destination are collectively referred to as multipath.

A conventional RF TX/RX (with one antenna on the TX side, and one antenna on the RX side) is referred to as Single Input Single Output (SISO) system; there is one stream of input digital data generated at the transmitter, resulting in one stream of digital output data demodulated at the receiver. The stream of digital data, however, then traverses all of the multipath between the transmitter and receiver. As the various paths have different path lengths, this causes delayed copies of the same signal to show up at the destination. Further, scatterers cause amplitude loss (attenuation) as well as phase changes in the scattered signals, so the various delayed copies arrive out-of-phase with each other and with different signal levels. All of the copies of the transmitted signal that arrive at the destination are summed in the receiving antenna; the resulting signal is therefore the vector sum of all the copies, and varies in amplitude and phase depending on the amplitudes and phases of the different multipath component. The received signal is therefore highly dependent on the scatterers in the environment; changing the obstacles in the environment, or moving the source and/or the destination within the same environment, causes the signal to change quite drastically.

Normally, scattering is a significant problem for radio links, particularly high-speed digital links. When the delays are much smaller than the duration of a single modulated symbol (the symbol period), the effect is to cause fading, or changes in the strength of the received signals. The scattered signals combine additively (in-phase) or subtractively (out-of-phase), resulting in large changes in amplitude as the receiver is moved over small distances or turned through small angles. When the delays are equal to or greater than a symbol duration, on the other hand, inter-symbol interference is also caused; a delayed copy of a previously transmitted symbol can overlay a subsequent symbol at the destination, making it difficult or impossible to recover the data. Techniques such as equalization and Rake receivers (sorting out the multipath using special filters), orthogonal frequency division multiplexing (OFDM) modulation (increasing the symbol period by spreading the signal over more carriers), and so on are used to defend against multipath and mitigate its effects. An additional defense against fading can be obtained with the use of diversity of various types (which is why most Access Points (APs) and some clients use two antennas rather than just one). However, all things considered, multipath reduces the channel capacity (information capacity) of the RF channel between a conventional RF transmitter and receiver, and is therefore the enemy of conventional digital modulation techniques.

MIMO turns this entire problem on its head. Rather than treating scattering and multipath as enemies, it instead takes advantage of the fact that multiple paths are available to convey information from a source to a destination, and actually puts the multipath effects to work in order to increase the effective channel capacity. In a highly oversimplified sense, each of the paths between the source and destination is treated as a separate, isolated RF channel, and a separate stream of information is sent down that channel. Thus, for example, if it is possible to identify and isolate four distinct and separate paths between a source and destination, it is possible to send four times the amount of information, merely by using each of the four paths to convey a different digital bitstream. Scattering is therefore put to work to increase the channel capacity, rather than being treated as a nuisance that reduces the capacity.

In actual fact, one does not simply identify the scatterers in the environment and "shoot" separate beams of RF energy at each one, so that they can reflect off the scatterers and arrive at the receiver. Instead, the channel is treated as a single entity having a set of independent propagation modes; a set of antennas is used at both transmitter and receiver to inject RF energy into each of these propagation modes, modulating the injected energy for each mode with a different digital bitstream. The transmit antennas are modulated with a set of RF signals derived from the data to be transmitted; the signals have different phases and amplitudes carefully calculated to produce a complex resultant signal at the receiver, after all the scatterers have done their work. The receiver picks up the resultant signals from each antenna, and decodes them back into a single (high bandwidth) data stream. Matrix mathematics with complex matrices are used at both transmitter and receiver to obtain the signal to be transmitted at each antenna, as well as to combine the signals picked up by the receive antennas. Mathematically, the product of the TX and RX matrices with the transfer function of the RF channel between transmitter and receiver creates a high-bandwidth data path.

9.1.2 Correlated vs. Uncorrelated Scattering

For MIMO to provide the bandwidth increase possible, some constraints must be placed on the environment, as described here.

The first constraint is that there must be scatterers in the environment that are visible to both transmitter and receiver. It is a little-known fact that in a clean, obstacle-free RF environment, MIMO actually fails to work, as the independent propagation modes of the RF channel disappear and the situation collapses down to a single-antenna, single-stream case. (Put another way, if there are no obstacles and hence no scatterers, there is no way to create multiple paths from transmitter to receiver, and hence no way to split up and send the information down different paths; thus there will be no increase in channel capacity.) MIMO therefore works best when there are a large number of metallic obstacles in the environment, causing many non-line-of-sight propagation paths, and producing what is termed as a "rich scattering environment". In terms of statistical propagation modeling, MIMO works best in a channel that has a Rayleigh distribution, without the dominant line-of-sight path that characterizes a Ricean channel (see Section 8.4.1).

An immediate deduction that may be made from the above is that a MIMO system cannot be tested in an anechoic chamber. An anechoic chamber tries to produce no scattering at all, and attempts to model pure line-of-sight paths between source and destination. Thus the capacity of a MIMO system placed in a well-designed anechoic chamber would be no more than that of a single-antenna transmit/receive system (i.e., the SISO case).

The second constraint is that the multipath must be uncorrelated, as illustrated below.

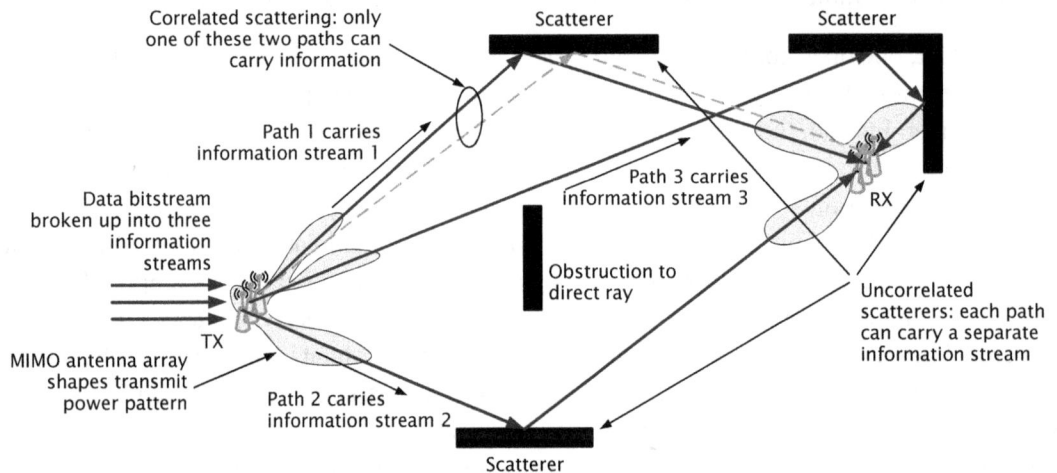

Figure 9.1: Multiple Uncorrelated Channels

Ideally, the multipath in a MIMO channel should occur with different scatterers, located at various large angles from each other (relative to the transmitter and receiver), and causing large angular spread in the received signals. It then becomes easy to distinguish the individual paths, and decorrelate them to extract the information corresponding to the different digital bitstreams. However, if the multipath is correlated – for example, all of the scattering occurs off a single very large metal sheet – then the individual signals traveling down the different paths become difficult or impossible to distinguish from each other, and no channel capacity multiplication is possible; all of the signals must carry the same information.

This means, for example, that it is impossible to test a MIMO system in a reverberation chamber. A reverberation chamber has an extremely rich multipath environment, but the multipath is all correlated (as is intended). Again, the capacity of the system would not exceed the SISO case. The same applies to a screened room.

Testing a MIMO system is therefore complicated by the need to provide an environment, real or simulated, that contains multiple uncorrelated scatterers. The number of uncorrelated scatterers corresponds to a statistical property of the channel referred to as the rank of the channel matrix. In order to support all the propagation modes required to carry the necessary information, the rank of the channel matrix must be greater than or equal to the number of transmit and receive streams.

9.1.3 Characterizing the Channel

A "channel" is basically an RF link between a transmitter and a receiver, having certain transmission properties (usually expressed as a complex function referred to as the channel transfer function). In an indoor environment that is rich in scatterers, the channel comprises the net effect of all of the multipath signals between the transmitter and receiver. As the multipath changes significantly from position to position, the channel properties correspondingly change for every pair of points within the environment. The properties of the channel are the main factor in how well a MIMO system works, and hence there is significant interest in determining them.

Determining the properties of the RF channel between any two points – that is determining the channel transfer function – is referred to as characterizing the channel. Once the channel is characterized, it is possible to accurately predict the received signal from a transmitted signal. In the case of MIMO, the channel is usually expressed as a matrix of complex numbers that represents the transfer functions of the various paths between the transmit and receive antennas; this is referred to as the channel matrix. Once the channel matrix is known, its rank can be calculated; as noted previously, the amount of information that can be carried by the channel can be determined from the rank of its channel matrix.

A channel can be characterized either by computing or modeling its properties (channel simulation) or by directly measuring its properties (channel sounding). The principles of channel simulation have already been covered in the last chapter and will not be repeated here. Channel sounding is done using a specialized piece of equipment known as a channel sounder, which transmits a complex RF signal into the channel, measures the received signal, and then uses these two signals to calculate the channel transfer function. Usually, channel sounding is done only for research into propagation, or when designing a new radio technology; this is because the equipment for channel sounding is quite complex and expensive. In the case of MIMO, however, channel sounding or estimation may be built into the WLAN radio in order to tune the transmitter and receiver signal processing to the properties of the channel between them, and thereby extract the maximum information carrying capacity of the channel; we will not consider these here.

Various types of channel sounders are in use today:

- Periodic pulse sounders; these operate in the time domain and directly measure the impulse response of the channel.

- Swept-frequency sounders, which operate in the frequency domain and measure the channel frequency response very much like a vector network analyzer or VNA (in fact, many sounders of this type use a VNA).

- Sliding correlator sounders; these use a spread-spectrum technique based on pseudonoise (PN) sequences to probe the multipath directly, and so in a sense are space-domain instruments.

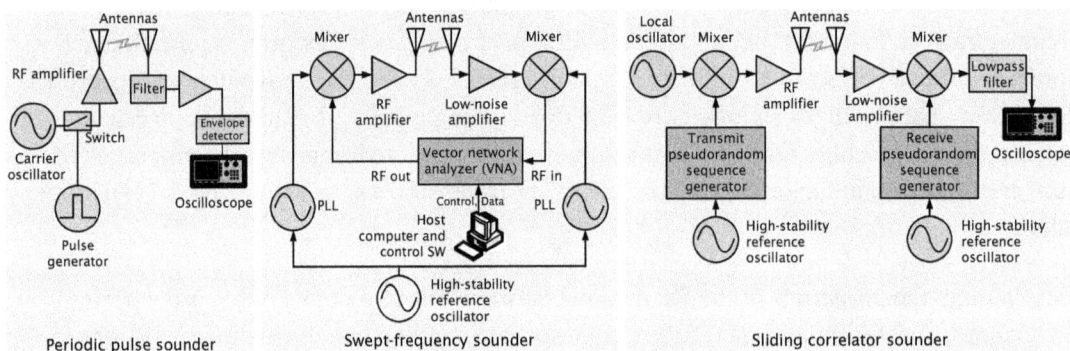

Figure 9.2: Channel Sounder Types

The periodic pulse channel sounder is conceptually simple, and relies on the fact that the channel impulse response can be measured by injecting a narrow pulse of energy into its input and measuring the signal received at its output. Note that this is also how a time-domain reflectometer (TDR) works; a pulse is sent down a transmission line, and the properties of the transmission line can be obtained by looking at the signal response. The frequency-domain properties of the channel can then be calculated by simply taking the inverse Fourier transform of the time-domain impulse response. The pulse should be as narrow as possible; an infinitely narrow pulse is required to measure the true impulse response, which is impossible, but a finite pulse will still give a close approximation over a narrower bandwidth. As shown in the figure above, this form of channel sounder is conceptually quite simple, comprising an RF pulse generator that modulates (switches on and off) a carrier wave, a pair of antennas, an amplifier, and an envelope detector. However, it is susceptible to interference (because the receiver must be very wideband) and it is difficult to construct a wideband system that can measure the short indoor propagation delays.

The swept-frequency channel sounder uses exactly the same principle as a standard VNA; in fact, in some cases an actual VNA is used as the heart of such a sounder. In essence, an RF sweep generator is used to generate RF signals over a range of frequencies; these are injected into the channel using an antenna, received with another antenna, amplified, and detected. As shown in the above figure, careful phase locking between the transmitter and receiver is essential to obtain the complex frequency response of the channel. This is a key issue with this type of channel sounder; while widely used, it is necessary to run RF cables between transmitter and receiver sites in order to implement the phase locked detection, which limits its range and usability.

The sliding correlator channel sounder measures the multipath signals directly. An RF signal (see the above figure) is modulated with a long pseudorandom digital sequence (i.e., a pseudonoise or PN sequence), and transmitted from a source antenna. At the receiving location, another antenna picks up the signal, which is then correlated against the same pseudorandom sequence. The correlation process basically slides the pseudorandom sequence in small time steps along the received signal; at each step, the received signal is multiplied by the pseudorandom sequence and the products are summed. Each multipath ray causes a correlation peak in the output, which can be displayed directly as the power-delay profile (PDP), or converted into the channel frequency response if desired. The sliding correlator channel sounder is inherently a spread-spectrum device, and therefore has the noise and interference rejection benefits of spread-spectrum technology.

9.1.4 MIMO Channel Capacity

The standard formula for calculating the capacity of a radio (or any information) channel is Shannon's limit on channel capacity:

$$C = B \times \log_2(1 + \text{SNR})$$

where C is the capacity in bits/second, B is the bandwidth in Hz, and SNR is the ratio of the signal power to the noise power (actually, noise + interference).

In the conventional (SISO) case, this directly gives the overall capacity of the system, as there is only a single effective channel between transmitter and receiver. A MIMO system, however, has M transmit antennas and N receive antennas, all simultaneously operating (N is assumed to be $>=M$ for systems that can realize the full capability of the channel). The N receive antennas, when used together, increase the signal-to-noise ratio (SNR) by N times, because the signals from multiple antennas when combined together causes a net increase in signal strength without a net increase in noise. The M transmit antennas, when used as a whole, with different signals being transmitted on each transmitted antenna, have two effects: the SNR is decreased by M times, because each separate transmitted signal interferes with the others at the receive antennas; however, the net bandwidth is increased M times, because M different digital signals are transmitted (effectively one per antenna). The end result is an increase in the channel capacity. The actual formula for the channel capacity is a complex function involving the channel matrix, but can be approximated as follows:

$$C = M \times B \times \log_2(1 + (N/M) \times \text{SNR}) \quad \text{for } N >= M$$

Note that if there are more transmit antennas than receive antennas (i.e., $M > N$), then some of the transmit antennas cannot be used to directly increase the channel capacity in the above equation. However, the extra antennas can be used to improve the overall SNR using beamforming, as will be described later.

In the most commonly analyzed case, $M = N$ (i.e., the number of transmit antennas is equal to the number of receive antennas). In this case, the channel capacity formula simplifies to:

$$C = M \times B \times \log_2(1 + \text{SNR})$$

That is, the channel capacity of a MIMO system with M antennas (all used for transmit and receive) is M times the capacity of a SISO system operating at the same channel bandwidth and SNR. This assumes a rich scattering environment.

In general, a MIMO system is characterized by the number of transmit and receive antennas used, as these directly indicate the channel capacity improvement that can be obtained. Thus a 4×4 MIMO system has 4 transmit and 4 receive antennas, and can provide up to 4 times the data rate of a SISO system. A 2×3 MIMO system, on the other hand, has 3 receive antennas but only 2 transmit antennas; it can provide twice the data rate of an equivalent SISO system, but the extra receive antenna can be used to offer additional robustness against noise and interference.

Note that the above equation only provides the theoretical capacity of a MIMO system. In order to actually realize this capacity in practice, it is necessary to resort to a number of complicated elements: complex coding methods (space–time coding, or STC), smart antennas, and digital modulation/demodulation processing with accurate estimates of the channel (channel estimation). In practice, one can only approach the capacity in the above equation, not fully realize it.

9.1.5 MIMO vs. Diversity

The technique of diversity (either receive or transmit) is often confused with MIMO processing. The two are quite different; in fact, a MIMO system may employ diversity as well for added gain in range when the SNR is too low for usable data rate gains.

MIMO, as has already been discussed, is a technique for utilizing multipath in such a way that not only is fading eliminated, but also the net capacity of the system is increased by employing multipath to form separate channels. Diversity, instead, is a technique for overcoming (rather than utilizing) multipath; specifically, for dealing with the fading effects of multipath, caused when different multipath signals interfere with each other constructively and destructively. Diversity is based on the fact that when several antennas are arranged within an environment at certain fixed distances from each other, fading can cause the signal at some antennas to decrease but will cause the signal at other antennas to increase. That is, some antennas may fail to provide enough power, but it is unlikely that all antennas taken together will fail. Diversity is therefore a probabilistic technique for combining signals from different antennas in order to combat spatial and temporal fading effects.

Diversity is traditionally divided into two types: antenna diversity and temporal diversity. Antenna diversity uses multiple antennas to combat the effect of spatial fading; these may all have the same polarization, but be placed at different locations (space diversity); use different polarizations (polarization diversity); or even use different types of radiation patterns (pattern diversity). Antenna diversity can be used at the receiver, the transmitter, or both the locations. Temporal diversity, on the other hand, uses coding or modulation methods to overcome time-varying fading. Examples of temporal diversity are: frequency diversity (e.g., changing channels when an existing channel fades out); code diversity (e.g., using spread-spectrum coding to decorrelate multipath signals); and time diversity (e.g., using error-correcting codes to recover from fades).

Diversity relies on the multipath producing multiple copies of the same signal at the receiver, but with different amplitudes at different times (or locations). The individual copies may fade in and out as conditions change, but it is much less likely that all of the copies will fade out at the same time; thus the aggregate signal can remain roughly constant even as the multipath components come and go.

Once multiple copies of the same signal are available (e.g., from multiple antennas in the case of space diversity), they are either selected or linearly combined, in order to produce an output signal that is much more immune to the effects of spatial and temporal fading due to multipath. Selecting the strongest signal is referred to as switched diversity combining, and is quite straightforward to implement at the receiving end: the signal with the highest signal strength is selected and the others are simply ignored. (The principle of reciprocity in RF systems means that switched diversity works the other way as well; once the antenna providing the strongest signal has been located, there is just as much benefit to using it for transmitting as for receiving.) Combining can be done by signal processing functions such as maximum ratio combining (MRC), which weights each copy of the signal and then adds the copies together in such a way as to maximize the SNR of the overall signal.

9.1.6 Beamforming

The IEEE 802.11n draft standard allows beamforming to be used as an optional mode, to enhance SNR. Beamforming is a process whereby multiple antennas are driven from a single transmit signal which is split and then modified in phase and amplitude in such a way that the radiation pattern resembles that of a directional antenna, and "beams" are formed in specific directions or propagation modes. That is, the antenna gain is increased in these directions. With N transmit antennas, it is possible to form $(N - 1)$ beams, each aimed in a different direction or propogation mode. The effect of forming these beams is that RF energy is increased in the desired direction(s) (e.g., towards the receiver) and decreased in undesired directions; the concentration of RF energy in the desired direction increases the signal level at the intended receiver, and the decrease of energy in undesired directions reduces interference to (and from) other devices. Beamforming has long been used in radar systems, where the technology is also known as a phased-array antenna. The process is illustrated below.

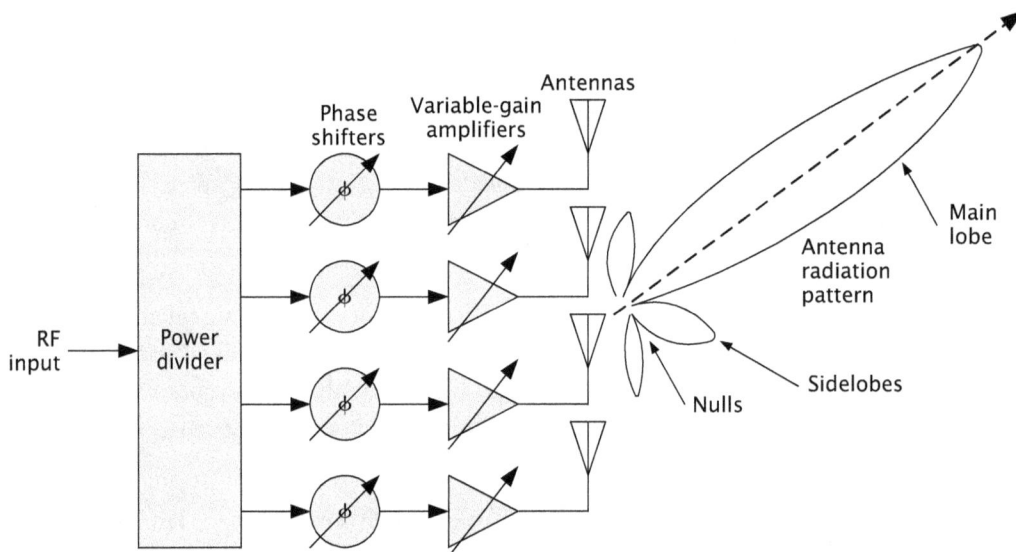

Figure 9.3: Beamforming

Beamforming can be profitably combined with MIMO transmission, especially when the number of transmit antennas exceeds the number of receive antennas. In this case, the number of spatial streams is limited by the number of receive antennas rather than the number of transmit antennas, and the available transmit antennas cannot be fully utilized to increase the data capacity of the system. For example, a transmitter with four antennas may be transmitting to a receiver with just two antennas; only two of the transmit antennas are actually required for data transfer, and only two spatial streams can be set up. However, the extra transmit antennas can be used to form beams and thereby focus more energy in the propagation modes that will actually reach the receiver. This results in a net increase in SNR at the receiver, causing either an increase in range (for the same traffic capacity and frame error ratio), or a reduction in frame error ratio (for the same range).

Beamforming is usually performed as part of the signal processing matrix arithmetic that is used to create the transmitted MIMO signals; it adds an extra step of complexity to the datapath, but does not require any additional components. Performing beamforming properly in an indoor environment requires a reasonably accurate knowledge of the channel (more specifically, the channel matrix). This is commonly obtained by having the transmitter and receiver exchange special "sounding" packets, which are essentially probe packets containing a known pattern that are transmitted in both directions and then analyzed. After the probe packets have been exchanged and analyzed, the transmitter can use the results of the analysis to construct an approximate channel matrix, which can then be used for beamforming in both the receive and transmit modes.

9.1.7 MIMO Coding Methods

The Shannon channel capacity equation given in Section 9.1.4 forms the keystone of communications and information theory, but it does not indicate how the theoretically available channel capacity can be realized in practice. The channel capacity equation merely indicates what can be achieved, not what will be achieved. Actually realizing the capacity of the channel is done by a different technique: coding.

"Coding" in this context refers to the process of encoding the information to be transferred in such a way that the available channel capacity is realized as fully as possible, given the constraints of signal level and noise. For example, if the channel has a relatively narrow bandwidth but a very low-noise level (high SNR), then it is possible to send a more complex signal, that is, transfer more bits per analog symbol – and thus take advantage of the higher SNR. On the other hand if the channel has a wide bandwidth but is plagued with high noise, then forward–error–correction (FEC) codes can be used; these allow recovery from errors induced by noise, but at the expense of bandwidth. Coding theory enables such tradeoffs to be made in order to come as close as possible to the theoretical channel capacity dictated by Shannon's equation.

Modern digital radios use a cascade of encoding and modulation methods to maximize their use of the channel capacity. For example, the OFDM transmitter in an IEEE 802.11a/g PHY first encodes the digital bits with a convolutional encoder to enable error recovery at the receiver, then interleaves the encoded output with an interleaver to spread noise bursts evenly across the message, then uses the result to modulate a set of subcarriers with Quadrature Amplitude Modulation (QAM) to set the ratio of bits per $4\,\mu s$ analog symbol, and finally upconverts all the subcarriers into the 16 MHz WLAN channel to fill the available channel bandwidth with information. A MIMO transmitter uses all of the above coding processes, but adds another layer – spatial coding. Spatial coding produces the signal that should be sent to each transmit antenna such that the required number of parallel and independent spatial streams can be generated and injected into the channel, and thereby obtains the performance gain of a MIMO system.

In its simplest form, spatial coding can consist merely of dividing the digital bitstream into M blocks, with each block having an equal number of bits, and transmitting these blocks (after suitable OFDM coding) on M different antennas. However, this does not realize the absolute maximum capacity from the channel. More complex coding methods, such as space–time block coding (STBC), can be used to achieve higher capacity under certain cases. These higher-level codes are options within the 802.11n draft standard.

9.1.8 Putting It All Together: The Layered Space–Time Architecture

One of the earliest MIMO prototypes was the Bell Labs Layered Space–Time Architecture, or BLAST, pioneered by Gerard Foschini at Bell Laboratories in 1996. (BLAST has spawned a tremendous number of subsequent variations in the research literature, such as V-BLAST, T-BLAST, D-BLAST, SD-BLAST, etc.) The BLAST architecture showed the first practical way of utilizing the multiple spatial streams realized by MIMO to transmit digital data. Since then, most MIMO implementations have followed this architecture or variations thereof. The following figure shows a simplified representation of BLAST, for a 4 × 4 MIMO system.

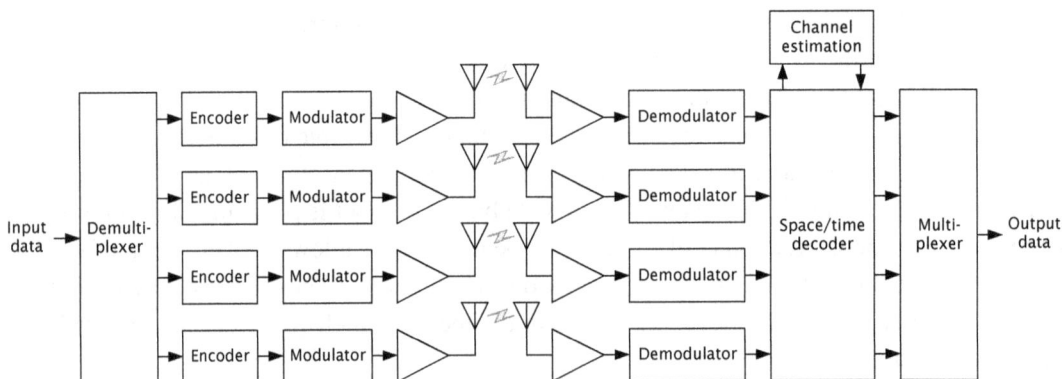

Figure 9.4: BLAST Architecture Overview

The BLAST architecture for an $N \times N$ MIMO system is relatively simple, once the various MIMO functions described in the foregoing sections have been understood. As shown in the figure, on the transmitter side the digital information bits to be sent are first divided (demultiplexed) into N streams, and then encoded suitably using the spatial coding block. The encoded streams are then modulated on to N separate transmitters and finally fed to N antennas, for transmission through the RF channel.

On the receiver side, N receive antennas pick up independent (uncorrelated) signals. These signals are subjected to receive signal processing to separate and extract the original N coded streams. The N recovered streams are passed through a decoder to convert them back into the original bits, which are finally multiplexed into a single digital information stream.

With this introduction to MIMO techniques in hand, we will look at the PHY layer described in the IEEE 802.11n draft standard. By necessity, the descriptions are kept brief; the reader is referred to the IEEE 802.11n draft document for more information.

9.2 The IEEE 802.11n PHY

The 802.11n PHY (digital radio) represents an ambitious and significant advance for 802.11 WLAN technology. It takes some of the most advanced techniques and concepts currently known in digital RF communications, and applies it to low-cost enterprise/consumer LANs. In doing so, it provides an order-of-magnitude increase in data rate (to as much as 600 Mb/s) while also promising to improve range and interference resistance.

9.2.1 Evolution of IEEE 802.11n

The IEEE 802.11n work started in the High Throughput Study Group in 2003, which was tasked with investigating the feasibility of a next-generation WLAN PHY that would offer at least 5–10 times the channel capacity of an 802.11g PHY. It was clear from the outset that MIMO would be a significant part of the solution, because the operating frequency ranges and channel bandwidths in the WLAN bands are defined by regulatory bodies and would hence be more-or-less the same as for standard 802.11a/g PHYs, but 5–10 times more bits had to be squeezed in somehow. After a good deal of preliminary work, the Study Group demonstrated that modern technology could yield a cost-effective and high-performance PHY capable of the desired level of bandwidth. As a result, the IEEE approved the IEEE 802.11n Project Authorization Request in September 2003, which resulted in the formation of the 802.11n Task Group.

The task group started by defining a set of functional requirements for the new PHY, a set of comparison criteria to be used in evaluating proposals, and a set of six channel models under which the proposed PHY was required to function. Various groups then brought in candidate proposals for the PHY standard; a total of four substantive proposals were made, which were then measured against the comparison criteria and each other. After a great deal of heated debate, the best features of the four proposals were combined into a single Joint Proposal, which was converted into the first draft of the 802.11n standard, published in March 2006. As of this writing the draft standard has been revised twice and is slowly making its way through the IEEE standards process, which involves numerous cycles of review, voting, and modification. The final 802.11n standard is not expected to be ratified until 2008 at the earliest; however, in the mean time, vendors are already creating chipsets and systems based on early drafts of the standard, in the hope that the final standard will not depart too far from the early versions.

9.2.2 Standard IEEE 802.11n Channel Models

To facilitate simulations, performance estimation, and comparisons of competing proposals, the IEEE 802.11n committee created a set of six channel models representing typical environments in which the 802.11n PHY is supposed to work. These channel models were derived from considerable industry and academic research work, and were specialized to deal with indoor MIMO propagation; they are based on the cluster models developed by Saleh and Valenzuela. The six models are labeled A through F, and are described in the following table:

Model	Represents
A	An ideal free-space environment with no scattering and simple flat fading
B	Residential (relatively small floor area and few metallic objects)
C	Small- to medium-size office area
D	Typical office area
E	Large office or indoor space, with many metallic objects
F	Large indoor or outdoor areas

These models were verified by extensive experimentation and data gathering. While the models were originally intended for simulation, design, and testing of competing 802.11n PHY proposals, the channel models are also very useful for the testing and performance measurement of actual 802.11n devices and systems.

9.2.3 IEEE 802.11n Operational Modes: Antennas, Bandwidth, and Coding

IEEE 802.11n is characterized by a tremendous variety of operating modes, which result from the large number of combinations of spatial streams, coding methods, modulation methods, and channel bandwidths. Currently over 300 different combinations of these parameters are possible; each combination results in a different PHY bit rate (and offers optimum performance for some combination of SNR and propagation environment). Each operating mode is identified by a number referred to as the Modulation Coding Scheme (MCS) index.

The various 802.11n PHY parameters are summarized in the table below:

Parameter	Description	Range of Values
Spatial streams	Number of independent spatial streams (equivalent to number of TX/RX antennas)	1, 2, 3, 4
R	Convolutional code density	1/2, 2/3, 3/4, 5/6
Modulation	Type of modulation applied to subcarrier	BPSK, QPSK, 16-QAM, 64-QAM
Channel bandwidth	Bandwidth of final transmitted signal	20, 40 MHz
GI	Guard interval between transmitted symbols	800, 400 ns

Note: There are several combinations where each of the 2, 3, or 4 spatial streams is given a different modulation type.

The various MCS indices are assigned to different combinations of the above parameters. These combinations produce PHY data rates ranging from 6.5 to 600 Mb/s, and everything in between. In general, for a 20 MHz bandwidth and 1 spatial stream (1 TX and 1 RX antenna), the maximum PHY data rate achievable is 72.2 Mb/s; for 2 spatial streams, 144.4 MHz; for 3 spatial streams, 216.7 Mb/s; and for the full 4 spatial streams (4 × 4 MIMO), 288.9 Mb/s. The PHY data rates more than double when the 40 MHz bandwidth is used, to a maximum of 600 Mb/s.

Note that the number of IEEE 802.11n options does not end with the above combinations. In addition to the normal convolutional codes, an optional Low-Density Parity Check (LDPC) coding is also possible, in cases where a stronger FEC is required. Also, in addition to the standard spatial multiplexing (i.e., one stream per subchannel), STBC is also supported as an option, as well as transmit beamforming (TBF).

9.2.4 Channel Estimation

To allow the receiver to properly decode the data in each frame, it is necessary to obtain an estimate of the channel properties between the transmitter and receiver, so that the matrix operations required to extract the data streams transported by each mode of the channel can be performed. Accurate channel estimation must be done frequently (preferably, prior to every frame) because the indoor channel is time varying, particularly in the case of mobile 802.11n stations. The 802.11n PHY achieves this by transmitting a special predefined sequence of signals referred to as the high-throughput Long Training Field (HT-LTF) symbols, in the preamble of each frame, before the actual medium access control (MAC) data is transmitted. As the HT-LTF is a known pattern, the receiver can use this to calculate and refine its channel estimates. The receiver effectively sets up a candidate channel matrix and then modifies it to cause the expected signal (the HT-LTF pattern) to match the signal actually received. The number of transmitted HT-LTF symbols increases with the number of spatial streams, as the accuracy of the channel estimation required increases.

A second mode of channel estimation is used for special situations, such as beamforming (see below). This comprises sending predefined frames called sounding PPDUs (PLCP Protocol Data Units) from the transmitter to the receiver. Again, as the receiver knows the contents of the sounding frames in advance, it can compare the received signals with the expected values and thus obtain a more accurate estimate of the channel, in terms of a channel matrix. This channel matrix may then be sent back from the receiver to the transmitter in a subsequent explicit feedback packet. Alternatively, the reciprocity of the channel (i.e., the fact that the RF channel has the same properties in either direction) can be used; the transmitting station can simply wait until it receives a corresponding sounding packet from the far end, at which point it can compute its own channel matrix.

9.2.5 Adaptive Beamforming

The 802.11n PHY specification allows beamforming to be optionally performed when the number of transmit antennas exceeds the number of spatial streams, and when the channel between the receiver and transmitter is known accurately enough by the transmitter to permit it to send most of the signal energy in directions that will benefit the receiver. Beamforming requires a knowledge of the channel, which is obtained implicitly (by analyzing the HT-LTF portions of frames received from the far end) or explicitly (by using sounding packets). In either case, once the channel matrix is known, the transmitter adjusts the RF signals sent to the transmit antennas in such a way as to maximize the power directed toward the receiver (Note that 802.11n uses 'eigenbeamforming' based on propagation modes, rather than forming actual beams.).

In addition to actively forming beams toward the receiver, the 802.11n draft standard also includes a scheme for preventing unintentional beamforming. This can occur if the data being transmitted down the various spatial streams inadvertently forms correlated patterns (i.e., similar data sequences) that are synchronized to each other; for example, a binary sequence such as "10101010. . ." will split among the antennas such that the signals emitted from all the antennas are phase aligned. In a situation where the transmitted signals from multiple antennas are coherent in amplitude and phase, the radiation pattern will form beams. This is much like the manner in which antenna arrays obtain their directive characteristics by feeding multiple antennas with phase-shifted copies of the same signal. Unlike intentional beamforming, however, the pattern of lobes and nulls may not be oriented in such a way as to maximize the effect at the receiver, and thus unintentional beamforming can be detrimental to the system.

To avoid unintentional beamforming, the IEEE 802.11n draft standard uses a process known as Cyclic Delay Diversity (CDD), which basically just offsets each spatial stream by a different constant, non-coherent delay. The offset considerably lowers the likelihood of correlated signals being transmitted by two or more antennas. This, in conjunction with a pseudorandom scrambler run over the transmitted data bits, ensures that the likelihood of two spatial streams correlating is very low.

9.2.6 The IEEE 802.11n Transmitter Datapath

The 802.11n transmitter is basically a superset of the standard 802.11a or 802.11g transmitter datapath; it consists of two or more sets of simultaneously operating OFDM datapaths with some special signal processing logic to implement the spatial multiplexing functions. A conceptual block diagram of a 4×4 MIMO transmitter system is shown below. Note that the same diagram can be extended to 3×3, 2×2, etc. by simply omitting channels.

Proceeding from left to right in the below figure:

    a.   The digital bitstream (i.e., the PHY layer convergence protocol (PLCP)-framed MAC data) is scrambled and then split into two streams.

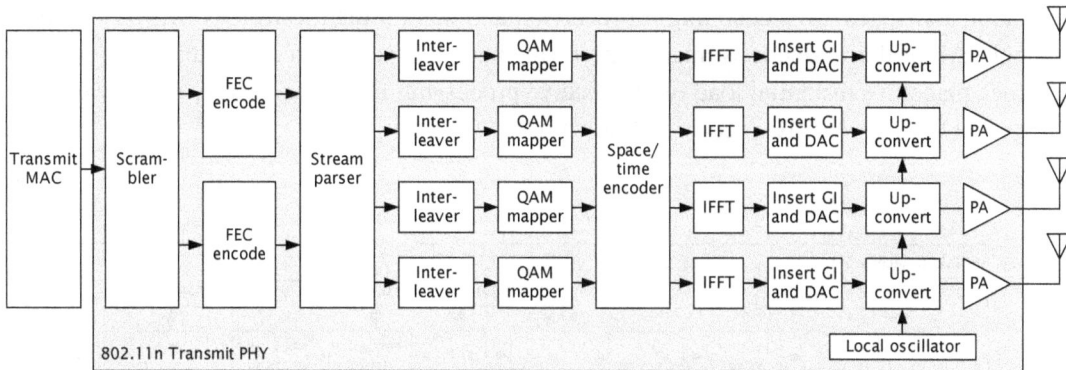

Figure 9.5: IEEE 802.11n Transmitter Architecture

b. Each stream is passed through a convolutional coder, that implements FEC coding on the digital bits.

c. The outputs of the convolutional coders are processed by a stream parser, to produce four streams of digital bits from the original single stream.

d. Each of the four streams is converted to the appropriate modulation format (BPSK, 64-QAM, etc.). Note that it is possible, in 802.11n, to have a different modulation format for each stream.

e. An optional STBC step is carried out on the four streams taken as a unit.

f. Spatial mapping, including beamforming and CDD, is then performed to produce four spatial data streams to be transmitted out to the four antennas.

g. The standard OFDM modulation process is carried out: the four streams are modulated on to the OFDM subcarriers using a set of inverse FFT blocks, after which another stage of CDD may be performed.

h. The final transmitted symbols are created by adding the guard interval (GI) between symbols, and then filtering the symbols through a suitable spectrum-shaping window.

i. Finally, the baseband signals produced thereby are upconverted to the appropriate RF channel, filtered, amplified, and transmitted.

As can be seen, in the simple case the 802.11n MIMO transmit datapath looks much like four parallel copies of an 802.11g OFDM (SISO) datapath, with some significant added functions: demultiplexing of the transmitted digital data into four streams, and MIMO-specific spatial processing such as CDD and spatial mapping.

9.2.7 The IEEE 802.11n Receiver Datapath

An 802.11n receiver is substantially more complex than the corresponding 802.11n transmitter. The receiver must perform not only the usual digital receive functions such as synchronization, automatic gain control (AGC), and demodulation, but also accurate channel estimation and space–time decoding for receiving and combining the various MIMO signal channels. All of these functions must be performed at extremely high data rates (up to 600 Mb/s), which places a substantial load on the receive processing functions. A very high-level view of a typical 802.11n MIMO receiver is shown in the following figure.

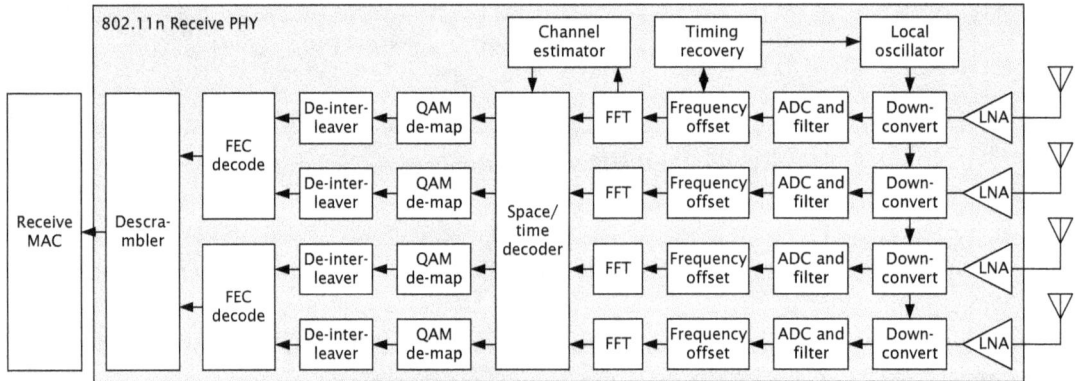

Figure 9.6: IEEE 802.11n Receiver Architecture

The receiver shown in the figure above comprises the following blocks:

a. A 4-channel downconverter and A/D converter is used to receive, amplify, filter, and mix down the RF signals to baseband or a low IF{aq expansion, and then convert them to digital form.

b. Carrier synchronization logic then locks on to the initial training sequences in the 802.11n frame and performs the final step of downconversion using internal Numerically Controlled Oscillators (NCOs), which are adjusted and synchronized to the training sequences.

c. A set of FFT blocks is used to convert the received symbols into the OFDM subcarriers.

d. A channel estimator block then uses the HT-LTF field (see above) and optional sounding packets to perform channel estimation, and obtain the channel matrix.

e. A space–time decoder block inverts the spatial mapping and STC that was originally performed at the transmitter, to produce the four modulated streams.

f. A BPSK/QPSK/QAM demodulator then converts the modulated streams into data bits.

g. A Viterbi decoder performs convolutional decoding and FEC processing on the data bits.

h. Finally, a de-interleaver and descrambler converts the four parallel bitstreams into a single interleaved, descrambled stream, which is output to the MAC logic.

9.3 A New PLCP/MAC Layer

IEEE 802.11n introduces a number of enhancements and extensions to the basic 802.11 MAC and PLCP formats. These extensions serve multiple purposes:

- Increasing the efficiency of data transfers to reduce per-frame overhead and thereby take advantage of the increased PHY data rates.

- Ensuring coexistence with legacy 802.11a/b/g devices.

- Provide support for channel sounding and estimation.

We will examine some of these extensions and their purposes in the following sections.

9.3.1 Three PLCP Formats

The PLCP is the term given to an outer framing header (and some simple protocol functions) applied to 802.11 MAC frames just prior to transmission on the physical medium. The PLCP frame header contains synchronization, channel estimation, modulation type, and frame length information fields, plus some protection bits to enable the receiver to verify that a received PLCP header is in fact valid. The receiver uses the PLCP header to lock to the incoming data, align, and set up its RF datapath (e.g., AGC parameters), and determine how to decode the actual MAC frame.

IEEE 802.11n currently specifies not one but three new PLCP formats. One format is referred to as "non-HT", and is basically the same as the standard 802.11a or 802.11g PLCP framing; it is used when operating as an 802.11a or 802.11g PHY. (In order to preserve interoperation with older devices, it is necessary for the 802.11n PHY to act as an 802.11a/g, 802.11b, and even an original 802.11 PHY, so that an 802.11n device can transmit to and receive from any legacy device.) The second format is referred to as "HT Mixed Mode"; it comprises a legacy 802.11a/g PLCP header immediately followed by the special 802.11n PLCP header. As the initial portion of the PLCP header is decodable by legacy devices, this PLCP format allows legacy 802.11a/g devices to detect when an 802.11n device is transmitting, and to stay off the air until the transmission is finished. Finally, the third format is called "HT Greenfield", and is used when only 802.11n devices are present; it contains only the special 802.11n PLCP header, and is not decodable by any legacy device.

The following figure depicts these three PLCP frame formats.

In the figure, the L-STF, L-LTF, and L-SIG fields in the non-HT and HT mixed-mode PLCP frames exactly correspond to the short training symbols, long training symbols, and SIGNAL fields of the standard 802.11a/g OFDM PLCP header:

- The L-STF is used for signal detection, AGC setting, diversity selection, coarse frequency tuning, and timing synchronization.

Non-HT (legacy) format

HT mixed-mode format

HT Greenfield format

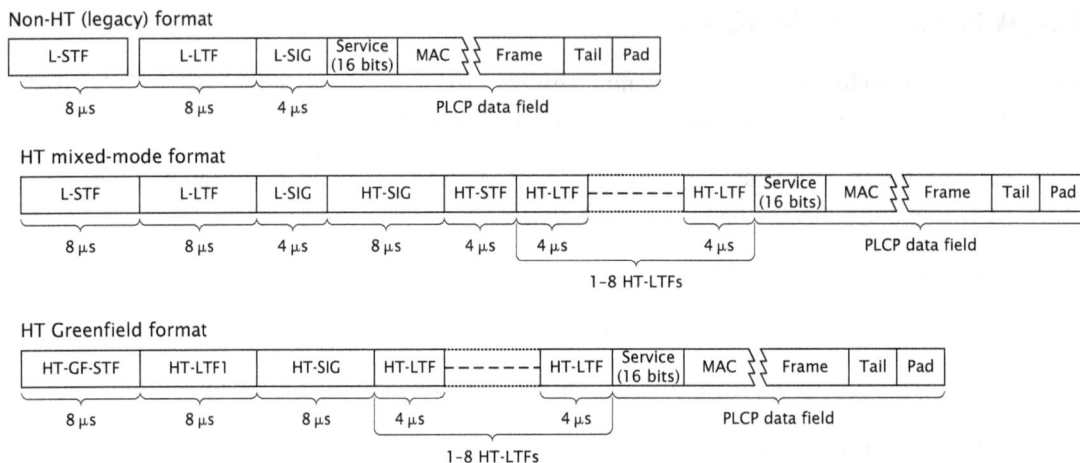

Figure 9.7: IEEE 802.11n PLCP Frame Formats

- The L-LTF is used for fine frequency offset estimation and tuning (i.e., centering the receiver's passband on the transmitted signal).

- The L-SIG specifies the PHY bit rate (equivalent to the modulation type) and the total length of the MAC frame in bytes, and is necessary in order to demodulate the remaining frame data.

In the HT mixed mode and Greenfield cases, the following fields are present:

- HT-LTF1, which is used for fine frequency offset tuning.

- HT-SIG, which specifies the modulation scheme used, as well as various options for the 802.11n PHY, and is used to decode the remainder of the frame.

- HT-STF (HT-GF-STF in Greenfield mode), which is used to improve AGC training, which in turn is essential for proper MIMO decoding.

- HT-LTF: multiple HT-LTF symbols are sent in order to allow the receiver to estimate the MIMO channel, as well as to perform additional channel sounding for use by optional modes such as beamforming or STBC.

9.3.2 Increasing Efficiency: Aggregation

IEEE 802.11n transmits packet payloads at a very high bit rate (up to 600 Mb/s). However, there is an issue with actually realizing this bit rate, and providing a high throughput to upper-layer protocols and user applications: the problem is that the amount of overhead involved with transmitting an 802.11 packet remains relatively constant even though the data rate has gone up by an order of magnitude, and so the efficiency drops sharply. In order to deliver a high throughput for user applications, it is necessary to increase efficiency.

The fixed overhead associated with each 802.11 frame involves:

- The PLCP header, which provides synchronization and channel estimation, and contains elements such as training fields that cannot be reduced without affecting the receiver.

- The gaps between packets (SIFS, DIFS, etc.), which allow the radios to switch between transmit and receive and also allows the channel to settle and noise to be estimated.

- The backoff intervals required for reducing collision probability in a multiple access situation.

- The acknowledgment packet (ACK) frames that must be sent to confirm delivery of the MAC frames.

- This overhead is dependent on physical properties (such as the acquisition time of the RF receiver) and the basic protocol, and cannot be eliminated or even significantly reduced.

In the case of IEEE 802.11n, the overhead can amount to over $200\,\mu s$ per packet, in Greenfield or mixed modes; most of this is taken up by the SIFS, DIFS, and backoff period. If a single 1500 byte frame (the maximum size that is usually transferred on the Ethernet infrastructure) is transmitted at 600 Mb/s, the 802.11n MAC frame requires only $20\,\mu s$ to transmit; however, the net time expended per packet including overhead is $220\,\mu s$, resulting in an efficiency of under 10%. Clearly there is little point in developing a complex, high-speed PHY if 90% of the speed improvements are lost due to protocol overhead.

In order to increase efficiency, the 802.11n PHY defines several features to allow multiple blocks of user data to be transmitted before the inter-frame gap and acknowledgment overhead must be paid. One of the key features is aggregation. Aggregation is done by concatenating several frames or user-level packets together into one much larger block, and sending the whole block as a single frame; the preamble, SIFS, DIFS, backoff, and ACK frame overhead is incurred only once for each frame, instead of once per user data block. This proportionally reduces the amount of overhead per frame, and enables much more of the available PHY bit rate to be realized for actual data transfer.

The 802.11n draft standard provides for two different types of aggregation, referred to as "A-MSDU" aggregation and "A-MPDU" aggregation. A-MSDU (Aggregated MAC Service Data Unit) aggregation is performed at the top of the MAC protocol layer (i.e., on user data blocks), while A-MPDU (Aggregated MAC Protocol Data Unit) aggregation is done at the bottom of the MAC layer, on MAC frames before they are encapsulated in a PLCP header and transmitted. The following figure depicts these two types of aggregated frames.

A-MPDU format

| MPDU delimiter | MPDU | Pad | MPDU delimiter | MPDU | Pad | - - - | MPDU delimiter | MPDU | Pad |

A-MPDU subframe 1   A-MPDU subframe 2   A-MPDU subframe n

| Reserved | MPDU length | CRC | Signature |

4 bits   12 bits   8 bits   8 bits

A-MSDU format

MAC payload field (frame body)

| MAC header | Subframe header | MSDU | Pad | Subframe header | MSDU | Pad | - - - | Subframe header | MSDU | Pad | MAC FCS |

A-MSDU subframe 1   A-MSDU subframe 2   A-MSDU subframe n

| DA | SA | Length |

48 bits   48 bits   16 bits

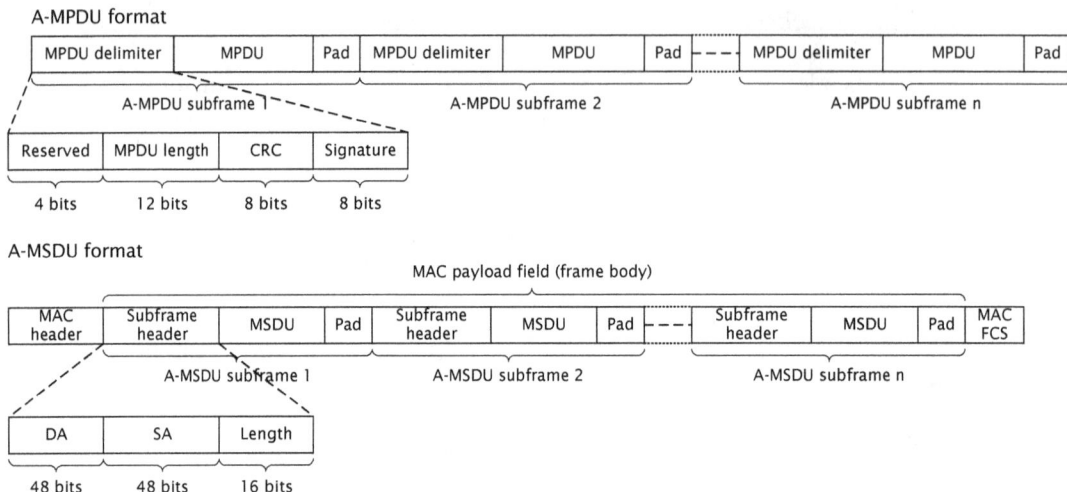

Figure 9.8: A-MPDU and A-MSDU Aggregated Frame Formats

Each subframe in the A-MSDU aggregate above can contain a payload of up to 2304 bytes, but the maximum size of the total aggregate cannot exceed 4095 bytes. An A-MPDU aggregate, however, can be up to 65,535 bytes in size. In either case, the amount of data that can be transferred before incurring overhead becomes much larger, as a result of aggregation. For example, transferring a full-size A-MPDU (64 KBs) results in increasing the efficiency to approximately 81% at a 600 Mb/s PHY data rate (from 10% without). Of course, the price of aggregation is complexity, both in the endstation and in the AP; these devices must now gather, buffer, and group frames prior to transmitting them.

The 802.11n draft also introduces the concept of a Reduced Inter-frame Spacing (RIFS) of 2 μs, which can be used when multiple consecutive frames are being originated from the same transmitter. Normally, an 802.11 data frame and the corresponding acknowledgment frame cannot be separated by any less than an SIFS (16 μs in the case of 802.11a, for example); this is a substantial amount of overhead, equivalent to almost a maximum-sized Ethernet frame at 600 Mb/s. The use of the RIFS can reduce the overhead considerably, further improving transfer efficiency. The downside, of course, is that due to limits on the transmit/ receive turnaround time, RIFS can only be used between consecutive frames from the same transmitter, with no intervening receive frame.

9.3.3 Quality of Service Extensions in 802.11n

One of the issues with wireless voice over IP (VoIP) handsets is that the current 802.11/802.11e power-save delivery mechanism is difficult to adapt to voice purposes, and is also somewhat inefficient when dealing with large numbers of handsets. The Power-Save

Multi-Poll protocol was devised in the 802.11n draft standard to deal with this issue. It is applicable to handsets using legacy PHY modes (e.g., 802.11b) as well as 802.11n handsets.

Essentially, the PSMP protocol allows an AP to transmit a PSMP frame that identifies a number of downlink (AP-to-handset) and uplink (handset-to-AP) slots during which data can be transferred. These slots are separated by an SIFS (or a RIFS, in the case of back-to-back transfers with an 802.11n PHY). As all other devices are required to wait for at least a DIFS before transmitting, once the medium is captured with a PSMP frame the AP and the power-save clients can retain the medium until all data is transferred. (This avoids the issue where non-power-save clients may intrude into the middle of a transfer to a power-save client, forcing the power-save client to stay awake for a longer period and thus expend more battery life.)

A PSMP frame is independent of the AP's beacons, and hence can be scheduled to occur at the expected voice packet interval rather than a fixed 100 ms beacon period. This solves a long-standing issue with conventional 802.11 unscheduled-delivery power-save methods, which rely on the beacon to signal the sleeping handset that buffered frames are available. The handsets may hence sleep most of the time, scheduling themselves to wake up at preset voice packet intervals; a PSMP frame will be transmitted by the AP to all the handsets at these intervals, enabling voice data to be efficiently transferred for a number of handsets before control goes back to the other devices trying to use the medium.

Another enhancement specified by 802.11n is the reverse direction exchange sequence. This enhancement is in view of the fact that a frame transmitted in one direction is very frequently followed by a frame transmitted in the reverse direction. For example, a TCP data segment sent to a device eventually produces a TCP acknowledgment segment in the reverse direction, and when the system has reached steady state every TCP data segment (or two) will be immediately followed by a TCP acknowledgment segment. The same is true for voice traffic; as voice traffic is bidirectional, a frame in one direction is predictably followed by a frame in the other direction.

Normally, the frames in either direction must separately contend for access to the medium, perform backoffs, incur different delays, etc.; all of this is both inefficient and error-prone. It would be preferable to allow a two-way frame exchange within a single medium access, which would not only increase efficiency but also reduce latency and jitter. Thus either device (client or AP) could contend for the medium once, paying the overhead required for contention at that time; once the medium had been acquired, they could rapidly exchange some predetermined number of frames before letting go of the medium.

The reverse direction exchange sequence thus allows a station that has seized the medium to provide a special Reverse Direction Grant (RDG) to its counterpart, which can then be used to transfer the return frame(s). The initiating (granting) station reserves the medium at the beginning of the sequence for the entire time required to transfer frames in both directions.

The frames are exchanged with inter-frame spacings of a SIFS to prevent other stations from getting into the middle of the reverse direction exchange. The result is that two-way transfer protocols with predictable patterns can be handled with much greater efficiency.

9.3.4 PHY Layer Support

As previously mentioned, the IEEE 802.11n protocol provides for special sounding, calibration, and channel estimation frames to be transmitted, in order to provide the facilities needed by the 802.11n PHY layer to operate at maximum efficiency. These frames are transmitted on behalf of the PHY layers, but are actually generated and received by the MAC layer. The reader is referred to the 802.11n draft standard for more details of these frames; they are rarely involved with test applications.

9.3.5 Legacy Interoperation

Successful networking technologies always have the burden of ensuring backwards compatibility, usually implying full interworking with all previously deployed equipments. The 802 LAN systems have been especially strong adherents to this rule; most 802 standards development groups try very hard to accommodate legacy devices when designing new protocols. For example, Ethernet interfaces have historically been able to transparently interoperate with all lower-speed versions; thus a 1000BASE-TX interface, which is capable of running at 1 Gb/s over twisted pair, can connect to and communicate with the legacy 100BASE-TX and 10BASE-T twisted-pair interfaces as well, automatically negotiating the best data transfer rate to use in each situation. IEEE 802.11n also has the same requirement; it must coexist with, and interoperate with, 802.11, 802.11b, and 802.11a/g stations that are operating on the same channel.

Legacy interoperation and coexistence in 802.11n is achieved mainly by proper selection of one of the three 802.11n PLCP headers. The PLCP headers contain fields which are capable of being received by legacy 802.11a/b/g devices, which interpret the data in them and thus detect that other devices are attempting to transmit. Further, 802.11n offers a mode in which an 802.11n PHY can communicate directly with an 802.11a/g PHY. The choice of which PLCP header is to be used is dependent on the composition of the basic service set of which the 802.11n device is a part, and is determined as follows:

- In situations where all devices are 802.11n, the Greenfield PLCP header is mandatory and sufficient, as coexistence with legacy devices is not necessary.

- In situations where some 802.11a/b/g devices exist, but the only requirement is that the 802.11n devices do not interfere with them, then the mixed-mode PLCP header is used; this header can be decoded by the legacy devices, and contains the information necessary to cause them to avoid interfering with the 802.11n devices.

- In situations where the 802.11n devices must actually communicate with the 802.11a/b/g legacy devices, the non-HT preamble is used; the entire frame format (not just the preamble) is comprehensible to legacy devices, and hence they can exchange data with 802.11n devices.

Note that 802.11n does not support a mode whereby it can communicate directly with an 802.11b CCK PHY (or original 802.11 DSSS PHY). To drop back to such modes, an 802.11b PHY has to be integrated into the 802.11n PHY.

9.4 The MIMO Testing Challenge

Adequate testing of 802.11n MIMO devices is much more complicated than the testing of similar 802.11a/b/g devices. This complexity arises from three issues:

a. The devices themselves are more complicated. Not only are there more antennas (and hence RF connections), but the protocol and PHY are more complex. Further, the adaptive nature of the MIMO PHY means that device behavior is far more complex.
b. Rather than trying to eliminate scattering effects, MIMO devices utilize them. Thus the effects of the environment must be factored in as part of the test setup for accurate results to be obtained. This is unlike 802.11a/b/g PHYs, where the environmental effects (apart from signal strength) could be largely ignored.
c. Coupling to the device under test (DUT) becomes quite complicated, especially with integral antennas. Simply placing the DUT in a chamber and using a probe antenna will not work. Direct cabled connections can serve for best-case performance measurements when $N \times N$ MIMO modes (i.e., 2×2, 3×3 and 4×4) are used, but even this is not feasible for an $M \times N$ mode (e.g., 2×3).

We will discuss these issues, and how they may be dealt with, in subsequent sections.

9.4.1 DUT Complexity

As should be clear from the foregoing, an 802.11n DUT is fundamentally more complex than anything that has been encountered in WLANs before. There are multiple transmitters, multiple receivers, and multiple antennas, and a great deal of complicated DSP. It is essential that signals from these multiple transmitters be processed by the test equipment, and signals be sent to the multiple receivers as well. This necessitates a test system that has multiple RF channels and a true MIMO baseband.

In addition, the 802.11n MAC is not only more complex but also operates at a much higher data rate. The test system therefore needs to be correspondingly more powerful. Overall, the complexity of a MIMO-based 802.11n tester is perhaps an order of magnitude more than the complexity of an equivalent 802.11a/b/g tester.

9.4.2 Performance and Spatial Effects

One of the key characteristics of MIMO devices is that they utilize the scattering properties of the environment to obtain their specified performance levels. It is possible to obtain a rather artificial view of the performance of the DUT by simply connecting cables between the antenna connections of the DUT and the RF ports of the tester. However, this method provides an unrealistic view of the DUT behavior; the performance measured thereby is unlikely to correspond to the performance in a real environment with actual scatterers. Further, it is common to find 802.11n devices with differing numbers of antennas on the endstation and AP sides; for example, a handheld device may have only two antennas, while the AP has three; or else a device with power consumption limitations may use three antennas on receive but only two antennas on transmit.

A much preferred method of testing is to interpose a scattering environment (either real, or realistically emulated) between the tester and the DUT. The properties of the environment then change the measured performance, and the results are much more likely to correspond to real-life behavior. An actual environment is extremely difficult to control and reproduce, however. Emulation is preferred, as an emulated environment can be easily reproduced at different sites. Emulation of the actual environment is done with a complex device called a channel emulator (or channel simulator), which mimics the properties of the RF channel between any two points in an environment, and thereby produces the same effect on the measured signals as a real environment. A channel emulator is a very useful device in MIMO testing, and will be discussed in more detail in a later section.

9.4.3 Coupling to the DUT

Coupling to a MIMO DUT, especially in an isolated (chambered or cabled) environment, is complicated by the fact that multiple antennas are used and therefore multiple RF paths exist between the tester and the DUT. As with any sophisticated RF system, isolation of these paths from one another by the test system is critical. This is particularly problematic for DUTs with built-in antennas, where direct connection to connectors on the DUT is usually ruled out, and the only option is to use external probes or test antennas. However, being a MIMO system, the coupling between the antennas is a key factor. This, for example, precludes simply placing such a DUT inside a chamber and then trying to feed it with several test antennas connected to the tester ports; the signals emitted by the test antennas will arrive with nearly equal strengths and multipath profiles at all of the DUT antennas; the channel will be fully correlated and true MIMO performance cannot be realized.

Figure 9.9 illustrates three possible modes of coupling a tester to a MIMO DUT.

The first alternative is feasible only in an open-air test environment; the actual scatterers and absorbers in the environment provide the MIMO channel effects. This is unfortunately highly

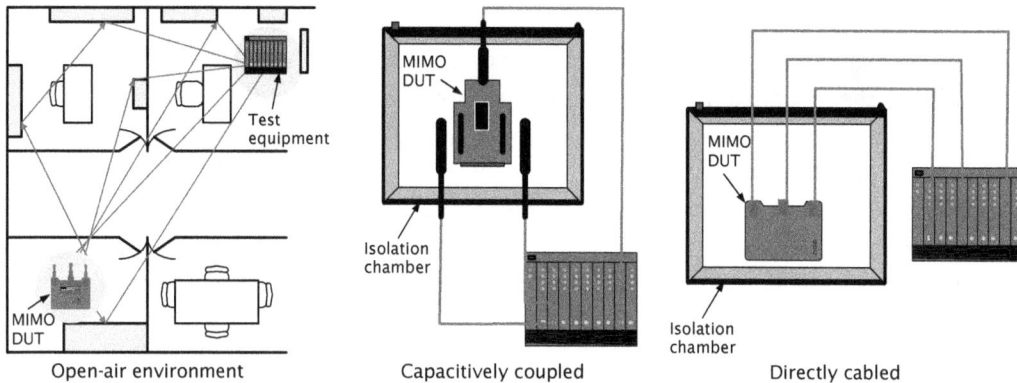

Figure 9.9: MIMO DUT Coupling Methods

limited – only one device can be tested in this environment without mutual interference – and not easily controlled. Further, the effects of the final deployment environment are difficult or impossible to reproduce in the test environment (actual scatterers have to be created and positioned), and thus the measured results may not correspond to the performance obtained in actual usage. Finally, the repeatability is likely to be poor unless special precautions are taken. However, this is by far the easiest environment to use for testing (for a home usage scenario, for instance, an actual home can be utilized as a test environment).

The second alternative, coupling via capacitive or inductive probes to the DUT antennas, is possible provided that it is feasible to provide enough isolation between the RF channels, either by shielding or by using a channel emulator (see below). This is in fact the only remotely viable alternative to open-air testing when the DUT has built-in antennas and it is not possible to bypass them. In this case, probes from the tester RF ports are placed in the reactive near field of the DUT (as close to the actual DUT antennas as possible) and coupling is mainly capacitive. Some limited shielding and isolation may be provided between the probes to reduce cross-coupling, but this is dependent on the mechanical construction of the DUT enclosure and the antenna separation, and is a significant problem.

The third possibility is the best approach, though it is only possible if the DUT has removable antennas, or otherwise provides direct access to its RF paths. Near-ideal isolation is achievable; the only significant leakage is within the DUT itself, which would in any event occur in actual usage as well. In this case, a channel emulator of some kind should be used to mimic the RF environment.

9.5 Channel Emulation

Once the properties of an RF channel have been measured and modeled, it can be simulated by means of a special device called a channel emulator. A channel emulator basically approximates the statistical properties of the desired RF channel – multipath, Doppler effects, time-varying behavior, noise, etc. – between one or more input ports and one or more output ports. (A SISO channel emulator has only one input port and one output port; a MIMO channel emulator, as can be expected, has multiple input ports and multiple output ports.) Bidirectional channel emulators enable the channel to be simulated in both directions, which is useful in most WLAN situations where devices exchange information rather than one device always transmitting and the other always receiving. MIMO channel emulators, besides having MIMO ports, can simulate the propagation effects between any pair of input and output ports, as well as coupling effects between output ports.

9.5.1 Realizing a Channel Emulator

Two primary approaches are used in the industry to actually implement channel emulators: RF analog methods and DSP.

9.5.2 Analog Channel Emulators

RF analog channel emulators use networks of phase shifters, delay lines, gain blocks, power splitters/combiners, and noise generators, all coupled together to emulate a predefined channel model. This is a very direct approach to modeling the channel, and the arrangement is not unlike the analog beamformers traditionally used with phased-array antennas. The functions of phase shifters and delay lines can be easily implemented using coaxial cables; a coaxial cable can create a constant delay that is nearly independent of frequency, and has a flat amplitude response and linear phase response over a wide bandwidth. Amplitude adjustments can be accomplished using either variable-gain amplifiers or (more simply) attenuators, either fixed or variable. The individual delay lines and attenuators are configured to simulate the individual multipath rays between a transmitter and receiver, as well as the phase shifts occurring at the points of reflection or diffraction; the attenuators simulate the path loss along the multipath rays. The splitters/combiners are used to tie together the various multipath simulation legs, as illustrated in the following figure for a 3 × 3 MIMO channel simulator.

This sort of "mechanical" emulation method has several benefits:

- It is inexpensive to build, consisting (in the extreme case) of purely passive components.

- It is inherently wideband and bidirectional, particularly when constructed from passive components. Even if it is constructed with active phase shifters and delay lines, the linear frequency range can be quite large.

- Relatively little noise and distortion is introduced into the system.

- It is directly mappable to time-domain channel models, which simplifies construction and understanding. For example, a standard power delay profile model of a channel can be directly mapped to the phase shifts and attenuations needed to model the channel.

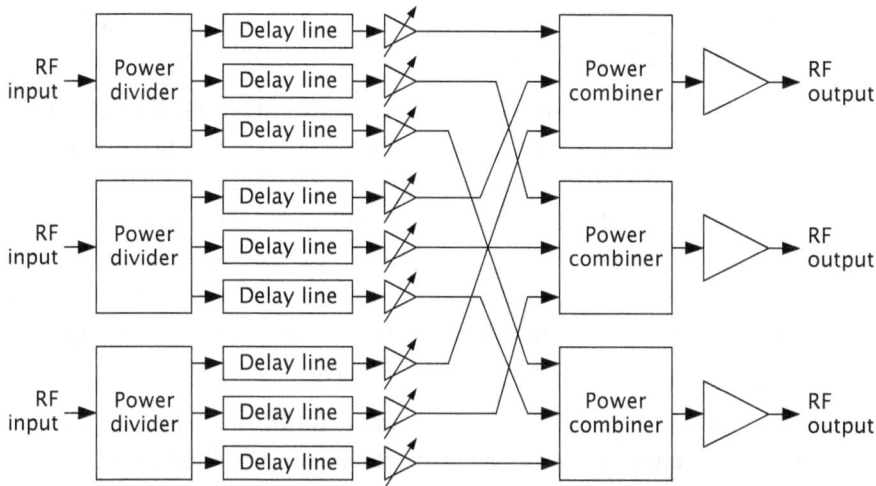

Figure 9.10: Analog Channel Emulation

However, there are also a number of significant disadvantages that make this arrangement much less attractive (and rarely used) in practice.

Firstly, the arrangement is mostly useful only for a single, fixed channel model. It is possible to use switched cables or tapped delay lines, but the complexity rapidly becomes unmanageable. Likewise, programmable phase shifters can be used, but the linearity and phase control range is problematic.

Secondly, this approach usually requires a lot of expertise to set up and configure. Besides requiring knowledge of the PDP of the channel, RF expertise is also needed to create and configure the system.

Finally, the range of emulation is limited by the leakage between devices and cables, as well as the port-to-port isolation of the power combiners. In general this approach is not recommended for more than a couple of phase shift taps and 2 ports of MIMO. Beyond that, the leakage of the resulting rats nest of cables and connections causes the results to be unpredictable.

For these reasons, analog channel emulators are infrequently used.

9.5.3 Digital Channel Emulators

A digital channel emulator uses DSP to simulate the effects of channel characteristics on RF signals. Essentially the channel H matrix is transformed to the time domain and then mapped to banks of Finite Impulse Response (FIR) filters, implemented digitally. Put another way, the channel impulse response is obtained and modeled directly in terms of filter coefficients.

The range (in terms of delay spread) and complexity (in terms of number of paths) of the channel to be emulated is limited only by the available signal processing power.

Each tap of an FIR filter represents a multipath signal in one direction, from transmitter to receiver. The system is thus basically unidirectional, so a bidirectional channel simulation necessitates a duplicate set of FIR filters in the opposite direction, usually (but not always) programmed with the identical channel model. The number of taps on each FIR filter therefore represents the maximum number of multipath rays that the system can emulate; the coefficients of the taps represent the attenuation and phase shifts that occur in each multipath ray. The maximum delay that can be introduced by the FIR filter directly corresponds to the maximum multipath propagation delay that can be simulated in the environment.

For example, a 4×4 MIMO channel emulator requires 16 FIR filters in either direction, as the paths from each transmit antenna to every receive antenna must be individually modeled. Each FIR filter may have 18 or more complex-valued taps, to simulate the amplitudes and phases of 18 different multipath runs. This represents a relatively large signal processing system.

To perform DSP on RF signals, it is necessary to perform A/D conversion first, then process the signals, then convert back to analog using a stage of D/A conversion. However, the input and output radio signals are usually in the RF/microwave domain (2.4 and 5 GHz for WLANs), and thus far beyond the capabilities of modern signal conversion and processing circuitry. On the other hand, the actual bandwidth of the system is quite limited; modeling a channel 83 MHz wide, for instance, is quite sufficient for the 2.4 GHz WLAN band. Thus downconversion is performed on the RF signals to transform them to a much lower frequency – sometimes even baseband – before channel emulation processing, followed by upconversion to restore the processed signals to the appropriate frequency band. This greatly reduces the requirements placed on the converters and DSP logic, and enables digital channel emulators to be realized with existing technologies and devices. Such an arrangement is represented in the following figure.

As shown in the figure, the RF input signals from the transmitter of the DUT or SUT (system under test) are first converted to a lower IF by a downconverter, then passed through the DSP logic, which implements the network of FIR filters required for the actual propagation modeling. If desired, a controlled level of Additive White Gaussian Noise (AWGN) is added to the processed signals to simulate the noise usually present on the actual channel. The result is converted back up to the original input frequency using an upconverter, and sent on to the receiver in the SUT or DUT.

Digital channel emulators are relatively narrowband, limited mainly by the A/D converters and the speed of the digital processing circuitry. They are also fairly expensive, due to the need for extremely linear and low-noise signal conversion, and the large amount of high-speed DSP employed. However, they are much more flexible and capable than their analog equivalents. Implementing different channel models involves merely changing the tap coefficients as

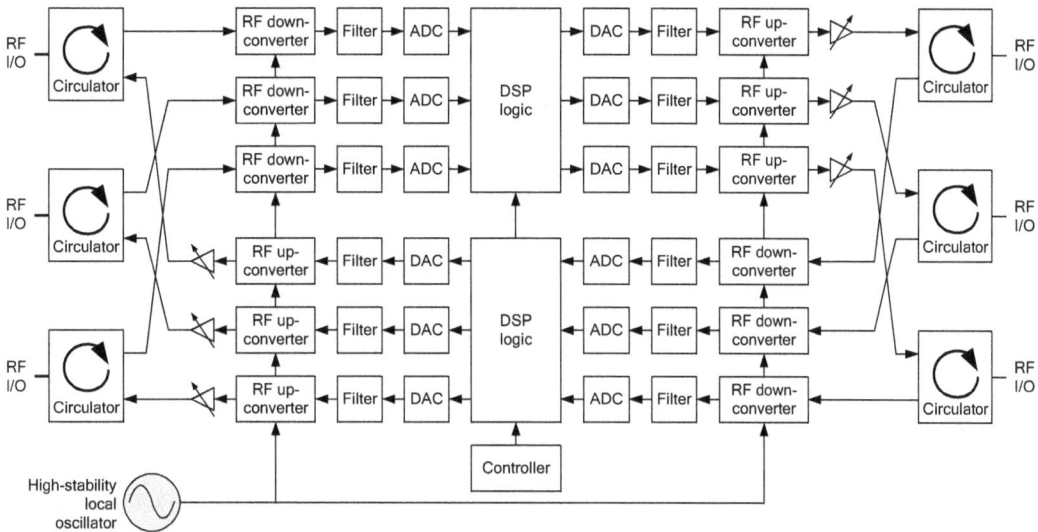

Figure 9.11: Digital Channel Emulation

well as adjusting the AWGN level, which can be done extremely quickly, sometimes on a packet-by-packet basis. High-density MIMO setups (4 × 4, or even 8 × 8) can be tested with ease, because there is no leakage or signal equalization to worry about. They are also usually accompanied by software packages with GUIs that greatly simplify the task of setting them up to emulate environments, and even come with predefined channel models for various applications such as 802.11n. Stand-alone versions of such channel emulators are available from Elektrobit, Spirent, and Azimuth Systems. Channel emulators built into performance test systems are also possible, for example in the VeriWave 802.11n test system.

9.6 Testing 802.11n MIMO Devices

Once the issues of emulating the channel and dealing with the more complicated radio and MAC are comprehended, testing an 802.11n system is not far different from testing a regular SISO 802.11a/b/g system. This section briefly covers some of the equipments and setups that may be used; all of the areas of testing discussed in the foregoing chapters, plus the general techniques used, apply to 802.11n performance, conformance, and functional test as well. Rather than repeat the details of the preceding chapters, a performance test setup for measuring the throughput and latency of an 802.11n device will be presented as an example, which can be easily extended to other applications.

9.6.1 Test Equipment

Test equipment for most 802.11n testing functions is very similar to that for 802.11a/b/g WLAN devices and systems, with the exception that multiple RF channels (and antennas) are involved. All of the basic tests and general approaches still apply (as they should, because the underlying technology and protocol has not changed; 802.11n represents a speed and complexity increase). Analog RF tests such as error vector magnitude, for instance, can be performed, after ensuring that the space–time encoding is controlled and accounted for. Performance tests at the MAC level or higher must be performed with all RF channels active, to obtain the maximum level of performance from the DUT; the test equipment must therefore be capable of communicating with the DUT in MIMO mode.

One key requirement for the test equipment, as already noted, is that some means of providing a controlled scattering environment (or a reasonably comprehensive simulation thereof, using a channel emulator) must be realized. This is perhaps the most complex part of performing MIMO testing.

9.6.2 Test Setups and Environments

An open-air test environment is at present the most common way of testing MIMO devices, principally because of simplicity and cost. (The cost of a stand-alone MIMO channel emulator is currently in the six-figure range.) Vendors have been known to lease or acquire entire buildings (offices, houses) in order to gain control over the nature of the open-air environment in which the testing is done. Failing that, a generic open-air lab or building environment is used for general equipment and performance checkout functions, as the results are not repeatable in such settings; cabled arrangements are then created for the more rigorous quality assurance tests. These cabled arrangements are usually no more than one-to-one connections of cables, through attenuators, between the DUT and the equipment performing the test function.

The test setup in an open-air environment is simple: the DUT is placed at one fixed location and the test equipment in another location, and packets passed between them. The "test equipment" is frequently only a complementary piece of off-the-shelf equipment running some representative application software; when testing an AP, for instance, a laptop client may be used. Obviously this is not by any means an optimal arrangement, but the WLAN 802.11n test industry is still in its infancy. This approach does have the benefit of being able to handle DUTs with embedded antennas, which characterizes many of the consumer-grade units being sold prior to the final ratification of the 802.11n draft standard.

A screened room or isolation chamber is not generally usable for MIMO testing, except in cases where simple connectivity and basic functionality are being checked, or firmware development and debug is being performed. In both situations the channel matrix degenerates to the SISO case: the screened room acts as a reverberation chamber, while an anechoic isolation chamber reduces or eliminates all the multipath entirely. The screened room is particularly egregious in this regard – the metallic walls reflect RF energy with little attenuation, and the repeated reflections around the room have been shown to produce multipath delay spreads that far exceed anything seen in the real world. Neither situation is useful for any kind of realistic performance test.

With a suitable channel emulator, either as a separate unit or built directly into the test equipment, it is possible to perform conducted tests on MIMO devices. (An isolation chamber is employed, as usual, to keep out unwanted interference.) Further, the channel emulator allows the number of antennas used by the DUT to differ from the number of antennas used by the test equipment. This enables, for example, testing the performance of a client device that has only two antennas communicating with an AP having three or four antennas. The antenna connectors of the DUT are simply hooked directly to the RF ports of the tester through the channel emulator. Unused RF ports of the channel emulator are terminated with 50 ohm terminators, and the signal processing internal to the emulator can then be configured to isolate (ignore) the unused ports. In some cases it may be useful to add a variable attenuator to each RF channel after the channel emulator, if the latter does not have enough attenuation range to fully emulate the flat fading characteristics over long distances in an indoor environment.

MIMO DUTs with embedded antennas that afford no means of direct cabled connections form a difficult problem for WLAN testing. These devices can obviously be tested in an open-air environment, but with the attendant problems of repeatability and reproducibility. As previously mentioned, a cabled setup requires some sort of capacitive coupling to the DUT antennas along with suitable isolation between channels. Internal probe points should be used if available.

9.6.3 An Example: Performance Test System

To illustrate the general nature of a 4 × 4 MIMO test system, a typical protocol test setup for functional testing on APs, as well as performance tests such as throughput, latency, forwarding rate, etc., is shown in the following figure.

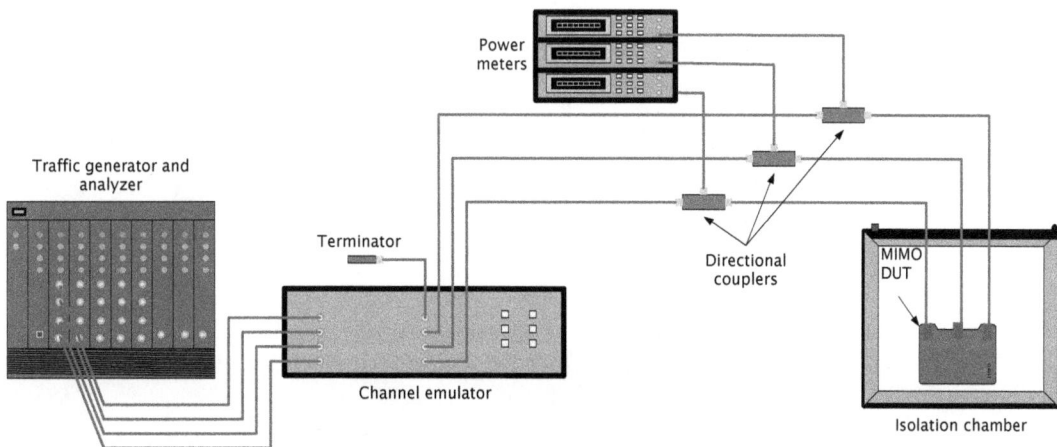

Figure 9.12: An 802.11n Performance Test System

From the figure, it is obvious that the system setup is quite straightforward. The traffic generator/analyzer is connected through a MIMO channel emulator to the antenna ports of the DUT. Four power meters are used to monitor the average RF power of the traffic stream injected into each port of the DUT. (The readings of the power meter are simply summed to determine the total received power at the DUT.) A shielded enclosure is used around the DUT and the cabled connections to ensure that external interference is excluded. The remainder of the test procedure is exactly according to the usual methods for conducting protocol tests, and has been described in the foregoing chapters. Note that if a traffic generator with embedded channel emulation is used, such as the equipment from VeriWave, then the external channel emulator is not necessary.

A Standards Guide

This Appendix provides a general guide to the regulatory and standards documents that govern WLAN equipment design and testing.

A.1 FCC Part 15

The Federal Communications Commission (FCC) Part 15 rules – more precisely, Title 47, Part 15, of the US Code of Federal Regulations – govern the operation of unlicensed RF emitters, categorized as unintentional radiators (i.e., devices that emit RF signals as a byproduct of their operation) and intentional radiators (i.e., devices that emit RF signals for the purpose of radio communications). To some extent the Part 15 rules could be considered to have spurred the development of Wireless LANs (WLANs) in the US when the FCC enabled the creation and sale of unlicensed spread-spectrum communications devices in the 2.4 and 5 GHz bands. The Part 15 rules are divided into six subparts:

- Subpart A lays down general rules for all types of unlicensed emitters, including measurement standards.

- Subpart B covers unintentional radiators, such as personal computers, TV sets, etc. This subpart is not applicable to WLANs.

- Subpart C regulates intentional radiators such as WLAN devices, cordless telephones, etc. Section 15.247 in this subpart specifically pertains to WLANs, and Sections 15.207 and 15.209 specify limits on spurious signals emitted outside the WLAN operating bands. Note that this subpart only covers WLAN devices in the 2.400 – 2.483 GHz ISM band.

- Subpart D describes unlicensed personal communications service devices and is not applicable to WLANs.

- Subpart E regulates unlicensed national information infrastructure (U-NII) devices, which includes WLANs operating in the 5.15, 5.25, and 5.725 GHz bands. Section 15.407(a) sets the operating power limits for these bands.

- Subpart F covers ultra-wideband (UWB) devices, such as ground-penetrating radars or newly developed wireless personal area network (WPAN) systems. This is also not applicable to WLANs.

Printed copies of the entire US Code of Federal Regulations containing the FCC Part 15 rules (and many other regulations) are available for purchase from the US Government Printing Office bookstore. However, an unofficial copy of the Part 15 rules in PDF format is available online at http://www.fcc.gov/oet/info/rules/part15/part15-8-14-06.pdf.

A.2 IEEE 802.11

The Institute of Electrical and Electronic Engineers (IEEE) 802.11 set of standards, as described in Chapter 1, forms the basis for wireless LAN technology worldwide. The most recent revision was published in 2007 as "IEEE 802.11-2007: Draft Standard for Information Technology – Telecommunications and information exchange between systems – Local and metropolitan area networks – Specific requirements – Part 11: Wireless LAN Medium Access Control (MAC) and Physical Layer (PHY) specifications".

The 802.11 standard contains a large and often confusing array of specifications, options, formal state diagrams, etc., with 19 subclauses and 16 annexes. However, the subclauses of most utility to the test engineer are:

- Clause 5: General description. Gives a good overview of the 802.11 architectural concepts and the functional model.

- Clause 7: Frame formats. Provides the detailed specification of the format of every 802.11 frame type (whether management, control, or data).

- Subclauses 9.1 and 9.2: MAC architecture and Distributed Coordination Function (DCF), respectively. This is the heart of the 802.11 standard as used today, and describes the methods used by the MAC to access the wireless medium.

- Subclauses 11.1–11.3: these describe some of the essential management functions performed by the MAC layer, such as authenticating and associating client stations.

- Clause 18: High rate direct sequence spread-spectrum PHY. This specifies the operating parameters of the 802.11b (1, 2, 5.5, and 11 Mb/s) PHY layer.

- Clause 17: Orthogonal frequency division multiplexing PHY. This specifies the 802.11a (5 GHz OFDM) PHY layer, and also serves as a base for the specification of the 802.11g (2.4 GHz OFDM) PHY.

- Clause 19: Extended Rate PHY (ERP) specification. This specifies the operation of the 802.11g PHY layer. Note that most of these specifications actually reference the nearly identical OFDM PHY specifications in Clause 17.

In addition to the above, Annex I (Regulatory classes) is also worth referencing, as it specifies the permissible transmit emissions in the 5 GHz band including the spectral masks and operating channels.

The upcoming IEEE 802.11n amendment, currently in revision 2 and still being balloted by the IEEE standards committees, will add another key clause to the above compendium (most likely Clause 20) as well as many changes and additions to the existing portions of Clauses 7, 9, and 11.

The IEEE 802.11.2 draft standard, which is still in the process of being prepared, gives recommendations for performance measurements on WLAN devices and systems. Many of the test procedures described in this book are codified in this draft standard. Unfortunately this document is not available as of the publication of this book; it is expected, however, that a draft will be available to the public shortly.

Ratified and published IEEE 802 standards are available for free download at http://standards. ieee.org/getieee802/802.11.html. IEEE 802 draft standards (including IEEE 802.11n) are also made available for purchase shortly after they are created, even though they may be subject to change and cannot be used for conformance purposes.

A.3 Wi-Fi® Alliance

The Wi-Fi® Alliance publishes a number of test plans for the formal certification of WLAN client and Access Point (AP) equipment, as well as specifications for selected areas of the 802.11 protocol, such as Quality of Service (QoS) and security. Their test plans are aimed at interoperability and conformance, rather than absolute performance (the Wi-Fi® Alliance after all being in the certification business rather than being concerned with benchmarking activities); however, many of the test procedures involve various kinds of performance measurements.

The list of Wi-Fi® Alliance test plans grows steadily as new technologies are introduced into the WLAN world. As of this writing, the following areas have been covered by test plans:

- Basic WLAN device functionality, particularly security features.

- QoS functions and multimedia support.

- Power-save functions.

- RF metrics (in conjunction with the CTIA; see below).

- Certifications for special devices such as Personal Digital Appliances (PDAs) and application-specific equipment such as printers and mobile phones.

Unfortunately, Wi-Fi® Alliance test plans are only available to members of the Wi-Fi® Alliance.

A.4 CTIA

The Cellular Telecommunications and Internet Association (CTIA) was founded in 1984 to represent all sectors of wireless communications in dialog with policy makers. It also operates an equipment testing and certification program, and has recently (in conjunction with the Wi-Fi® Alliance) published a test plan for the RF performance evaluation of mobile devices containing 802.11 WLAN radios. While this test plan pertains mainly to "converged" mobile devices (i.e., handheld devices containing a mix of cellular, WLAN, and other radio technologies), the techniques it describes are of significant interest for measurement and evaluation of pure WLAN equipment as well. For example, the Total radiated power (TRP) and Total isotropic sensitivity (TIS) metrics described in Chapter 4 are standardized in this document.

The CTIA test plan is available for free download at: http://files.ctia. org/pdf/CTIA_RFPerformanceEvalWiFiMobile_TestPlan_1_0.pdf

A.5 IETF BMWG

Performance measurement standardization at the IETF (Internet Engineering Task Force) goes back nearly 20 years, to the formation of the IETF Benchmarking Working Group (BMWG) in 1991 as a consequence of Scott Bradner's work at the Harvard University Network Device Test Lab (NDTL). The BMWG is mainly concerned with equipment performance measurements at Layer 3 and above (i.e., anything that directly concerns TCP/IP network equipment). However, its standards documents, published in the form of IETF RFCs, are referenced in many other areas of network measurement; for example, several of the metrics in the IEEE 802.11.2 draft standard trace their ancestry to BMWG work.

As of the writing of this book, about 20 performance measurement standards documents have been published by the BMWG, in areas ranging from the very basic (e.g., RFC 1242 on "Benchmarking Terminology for Network Interconnection Devices") to highly specialized (such as RFC 4098, "Terminology for Benchmarking BGP Device Convergence in the Control Plane"). BMWG documents are freely available for download online at http://www.ietf.org/ html.charters/bmwg-charter.html as well as any public repository of IETF documents.

In addition to specific performance measurement metrics and procedures, the IETF also publishes standards that define metrics or test procedures that are required as part of protocol implementations. For example, RFC 1889 (the Real Time Protocol standard) contains the standard procedure for computing smoothed interarrival jitter in voice applications, while RFC 4445 specifies a video quality metric known as the Media Delivery Index.

Selected Bibliography

The following bibliographic references were found useful when writing this book, and can serve as starting points for further study on any of the topics covered in the preceding chapters.

Agilent Application Note 1380-2 (document 5988-5411EN), "IEEE 802.11 Wireless LAN PHY Layer (RF) Operation and Measurement", Agilent Technologies Inc., 2002.

Agilent Application Note (document 5988-6788EN), "802.11a/g Manufacturing Test Application Note – A Guide to Getting Started", Agilent Technologies Inc., 2003.

Agilent Application Note (document 5988-9803EN), "Improve the Circuit Evaluation Efficiency of Wireless LAN Chipset Design with the ENA Series Network Analyzer", Agilent Technologies Inc., 2003.

Agilent Application Note (document 5989-3144EN), "8 Hints for Making and Interpreting EVM Measurements", Agilent Technologies Inc., 2005.

Agilent White Paper (document 5988-8155EN), "Using Agilent Wireless Protocol Test Solutions", Agilent Technologies Inc., 2002.

Alexander, B. and Snow, S. Preparing for wireless LANs: secrets to successful wireless deployment. *Cisco Packet Magazine*, 14(2), pp 36–39, 2002,

ANSI C63.4-2004, "Methods of Measurement of Radio-Noise Emissions from Low-Voltage Electrical and Electronic Equipment in the Range of 9 kHz to 40 GHz", 2003.

ANSI C63.5-2004, "Radiated Emission Measurements in Electromagnetic Interference (EMI) Control – Calibration of Antennas (9kHz to 40 GHz)", 2004.

Black, Uyless *Voice Over IP*. Prentice Hall, Upper Saddle River, New Jersey, 2000.

Carr, J.J. *Practical Radio Frequency Test & Measurement: A Technician's Handbook*. Elsevier, Woburn, MA, 1999.

CTIA, "Test Plan for RF Performance Evaluation of Wi-Fi Mobile Converged Devices", August 2006.

Dobkin, Daniel. *RF Engineering for Wireless Networks: Hardware, Antennas and Propagation*. Elsevier, Burlington, MA, 2005.

Durgin, Gregory. *Space-Time Wireless Channels*. Prentice Hall, Upper Saddle River, NJ, 2003.

Edney, J. and Arbaugh, W. *Real 802.11 Security: WiFi Protected Access and 802.11i*. Addison Wesley, Boston, MA, 2004.

Ekahau White Paper, "The A, B, Gs of Wi-Fi Site Surveys", Ekahau Inc., May 2004.

Farpoint Group White Paper FPG 2003-201.1, "Beyond the Site Survey: RF Spectrum Management for Wireless LANs", Farpoint Group, September 2003.

FCC 47CFR§15, "Title 47 of the Code of Federal Regulations, Federal Communications Commisson – Part 15 – Radio Frequency Devices", 2007.

Fluke Networks White Paper, (document 2564449), "Wireless Site Survey Best Practices", Fluke Networks Inc., 2006.

Foschini, G. Layered space-time architecture for wireless communication in a fading environment when using multi-element antennas, *Bell Labs Technical Journal*, pp 41–59, Autumn 1996.

Held, Gil. *Data Over Wireless Networks*. McGraw Hill, New York, NY, 2001.

Henry L. Bertoni. *Radio Propagation for Modern Wireless Systems*. Prentice Hall, Upper Saddle River, NJ, 2000.

IEC CISPR 16-1-1, "Specification for Radio Disturbance and Immunity Measuring Apparatus and Methods – Part 1-1: Radio Disturbance and Immunity Measuring Apparatus – Measuring Apparatus", November 2006.

IEC CISPR 22, "Information Technology Equipment – Radio Disturbance Characteristics – Limits and Methods of Measurement", 2005

IEEE Std 145-R2004, "Standard Definitions of Terms for Antennas", 2004.

IEEE Std 211-2003, "Standard Definition of Terms for Radio Wave Propagation", 2003.

IEEE Std 299-1997, "Standard Method for Measuring the Effectiveness of Electromagnetic Shielding Enclosures", 1997.

IEEE Std 802.11-2007, "Standard for Information Technology – Telecommunications and Information Exchange Between Systems – Local and Metropolitan Area Networks – Specific Requirements – Part 11: Wireless LAN Medium Access Control (MAC) and Physical Layer (PHY) Specifications", 2007.

IEEE P802.11n/D2.00, "Standard for Information Technology – Telecommunications and Information Exchange Between Systems – Local and Metropolitan Area Networks – Specific Requirements – Part 11: Wireless LAN Medium Access Control (MAC) and Physical Layer (PHY) Specifications: Amendment: Enhancements for Higher Throughput", February 2007.

IETF RFC 1242, "Benchmarking Terminology for Network Interconnection Devices", July 1991.

IETF RFC 1889, "RTP: A Transport Protocol for Real-Time Applications", January 1996.

IETF RFC 2285, "Benchmarking Terminology for LAN Switching Devices", February 1998.

IETF RFC 2432, "Terminology for IP Multicast Benchmarking", October 1998.

IETF RFC 2544, "Benchmarking Methodology for Network Interconnect Devices", March 1999.

IETF RFC 2889, "Benchmarking Methodology for LAN Switching Devices", August 2000.

IETF RFC 3550, "RTP: A Transport Protocol for Real-Time Applications", July 2003.

IETF RFC 3918, "Methodology for IP Multicast Benchmarking", October 2004.

INTF RFC 4445, "A Proposed Media Delivery Index (MDI)", April 2006.

IETF RFC 4814, "Hash and Stuffing: Overlooked Factors in Network Device Benchmarking", March 2007.

Intel White Paper, "Wireless Hotspot Deployment Guide", Intel Corporation, November 2003.

ITU-T Recommendation G. 107, "The E-Model, A Computational Model for Use in Transmission Planning", March 2003.

ITU-T Recommendation J.144-2004, "Objective Perceptual Video Quality Measurement Techniques for Digital Cable Television in the Presence of a Full Reference", February 2004.

Kraus and Marhefka, *Antennas For All Applications*, 3rd ed. McGraw Hill, New York, NY, 2003.

NIST Technical Note 1297, "Guidelines for Evaluating and Expressing the Uncertainty of NIST Measurement Results", 1994.

NIST Special Publication 672, "Experimentation and Measurement", U.S. Department of Commerce, reprinted May 1997.

Okamoto, Garret. *Smart Antenna Systems and Wireless LANs*. Kluwer Academic, Norwell, MA, 1999.

Pattan, Bruno. *Robust Modulation Methods and Smart Antennas in Wireless Communications*. Prentice Hall, Upper Saddle River, NJ, 1999.

Paulraj, Nabar. and Gore. *Introduction to Space-Time Wireless Communications*. Cambridge University Press, Cambridge, United Kingdom, 2003.

Petrat, H. 802.11a Measurement techniques and network issues. *Microwave Product Digest*, November 2003.

Rappaport, Theodore. *Wireless Communications: Principles and Practice*, 2nd ed. Prentice Hall, Upper Saddle River, NJ, 2002.

Sanchez, G. and Connor, P. How much is a dB worth?. *IEEE EMC Society Summer 2002 Newsletter*, pp 1–5.

Saunders, Simon. *Antennas and Propagation for Wireless Communication Systems*. John Wiley and Sons, West Sussex, England, 1999.

Trapeze Networks White Paper WP-PMW-501, "Planning and Managing Wireless LANs", Trapeze Networks Inc., 2004.

VQEG, "Final Report from the Video Quality Experts Group on the Validation of Objective Models of Video Quality Assessment", March 2000.

Z Technology Application Note #1003-1, "Measurement Antenna Selection and Use", Z Technology Inc., 2006

Index

www.ingramcontent.com/pod-product-compliance
Lightning Source LLC
Chambersburg PA
CBHW061356210326
41598CB00035B/6007